W9-ADW-718

TH
5618
.B34
1986

Ball, John E

    Tools, steel square,
joinery

FEB   1993

KALAMAZOO VALLEY COMMUNITY COLLEGE
LEARNING RESOURCES CENTER
KALAMAZOO, MICHIGAN   49009

65774

TH
5618
.B34
1986

AUDEL®

Carpenters and Builders Library Volume One

by John E. Ball

Revised and Edited by Tom Philbin

Macmillan Publishing Company
New York

Collier Macmillan Publishers
London

WITHDRAWN
COMMUNITY COLLEGE
LIBRARY

65774

FEB 16 1988

FIFTH EDITION

Copyright © 1965, 1970, 1976 by Howard W. Sams & Co., Inc.
Copyright © 1983 by The Bobbs-Merrill Co., Inc.
Copyright © 1986 by Macmillan Publishing Company, a division of Macmillan, Inc.

All rights reserved. No part of this book may be reproduced or transmitted in any form or by any means, electronic or mechanical, including photocopying, recording or by any information storage and retrieval system, without permission in writing from the Publisher.

While every precaution has been taken in the preparation of this book, the Publisher assumes no responsibility for errors or omissions. Neither is any liability assumed for damages resulting from the use of the information contained herein.

Macmillan Publishing Company
866 Third Avenue, New York, N.Y. 10022
Collier Macmillan Canada, Inc.

Library of Congress Cataloging-in-Publication Data
Ball, John E.
   Tools, steel square, joinery.

   (Carpenters and builders library / by John E. Ball; v. 1)
   Includes index.
   1. Carpentry—Tools. 2. Joinery.
I. Philbin, Tom, 1934–   . II. Title.
III. Series: Ball, John E. Carpenters and builders library; v. 1.
TH5618.B34 1982   694   82-1340
ISBN 0-672-23365-7   AACR2

Macmillan books are available at special discounts for bulk purchases for sales promotions, premiums, fund-raising, or educational use. For details, contact:
      Special Sales Director
      Macmillan Publishing Company
      866 Third Avenue
      New York, N.Y. 10022

10 9 8 7 6 5 4 3 2 1

Printed in the United States of America

Foreword

This book is designed to help the carpenter, the apprentice, and the do-it-yourselfer. This volume, first in a revised four-book series, embraces a wide variety of knowledge, from how to select wood to construction-style carpentry to furniture and cabinetmaking.

In today's carpentry, power tools—both portable and stationary—play an important role. But hand tools are still vital, and the accent in this book is on them. It covers a vast array of tools, from such common ones as the hammer, screwdriver, and wrench to such little-known but at times needed ones as inside and outside calipers. And the accent within these sections is on how to use the tools. Toward this end a large number of illustrations supplement the text.

While the information supplied is timely, there is also consideration of techniques that have long been a part of carpentry but are not known as well as they once were: such things as unusual tools, old-time joints, and processes like patternmaking. Here they are presented, the idea being that the carpenter should have a full grasp of old and new aspects of his trade so that he can deal with whatever comes along, whether it be erecting a modern building or repairing an antique.

No one book, of course, can hope to impart all the knowledge of a particular trade—no four books could. But it is hoped that this volume will give you a basic grasp of a fine trade and encourage further exploration.

Contents

CHAPTER 1

WOODS.. 9

Classification—growth—defects—seasoning—properties—pre-
servation—decay—plastic laminates—wood products—
summary—review questions

CHAPTER 2

NAILS 25

History—sizes—classification—tables—holding power—
selection—driving nails—summary—review questions

CHAPTER 3

SCREWS.. 45

Classification—length—head shape—point shape—strength
of screws—safe loads for screws—summary—review questions

CHAPTER 4

BOLTS 57

Classification—proportions—strength—thread—tables—summary—review questions

CHAPTER 5

THE WORKBENCH 67

Construction—bench attachments—tool storage—vises—
summary—review questions

CHAPTER 6

CARPENTERS' TOOLS.. 75

Selection—types and classification of tools—summary—
review questions

CHAPTER 7

GUIDING AND TESTING TOOLS 79

Straightedge—try square—miter square—framing square—
combination square—miter box—level—plumb bob—
summary—review questions

CHAPTER 8

MARKING TOOLS... 99

Classification—chalk line—carpenter's pencil—ordinary
pencil—scratch awl—scriber—compass—dividers—
summary—review questions

CHAPTER 9

MEASURING TOOLS...107

Carpenter's rule—folding rules—rules with attachments—measuring tips—marking gauges—summary—review questions

CHAPTER 10

HOLDING TOOLS..117

Horses—trestles—clamps—vises—summary—review questions

CHAPTER 11

TOOTHED CUTTING TOOLS 127

Saws—saw teeth—setting the saw—crosscut and ripsaws—action of ripsaw—action of crosscut saw—files—rasps—summary—review questions

CHAPTER 12

SHARPENING SAWS...139

Jointing—shaping—setting—filing—crosscut saws—ripsaws—dressing—circular saws—band saws—summary—review questions

CHAPTER 13

SHARP-EDGED CUTTING TOOLS......................................149

Paring chisel—firmer chisel—tang and socket chisels—butt, pocket, and mill chisels—gouge—chisel selection and use—taking care of chisels—drawknives—summary—review questions

CHAPTER 14

AXES AND HATCHETS 161

Various types of hatchets—axe—adze—summary—review questions

CHAPTER 15

SMOOTH FACING TOOLS 167

Spokeshave—jack plane—fore plane—jointer plane—smoothing plane—block plane—rabbet plane—router—plane irons—planing—scrapers—summary— review questions

CHAPTER 16

BORING TOOLS 189

Brad awls—gimlets— augers—twist drills—brace and bit—countersinks—reamers—summary—review questions

CHAPTER 17

FASTENING TOOLS 201

Hammers—screwdrivers—wrenches—summary—review questions

CHAPTER 18

SHARPENING TOOLS 215

Grinding wheels—tool grinders—oilstones—summary—review questions

CHAPTER 19

HOW TO SHARPEN TOOLS 219

Grinding—honing—use of oilstone—summary—review questions

CHAPTER 20

HOW TO USE THE STEEL SQUARE....................227

Applications—scales and graduations—scale problems—square-and-bevel problems—tables—rafter calculations—polygon cuts—brace measure—Essex board measure—summary—review questions

CHAPTER 21

JOINTS AND JOINERY................................269

Straight—dowel-pin—square corner—mitered—splice—rabbeted—scarf—mortise-and-tenon—dovetail—tongue-and-groove joints—summary—review questions

CHAPTER 22

CABINETMAKING JOINTS 299

Tools—glued joints—beveled joints—plowed-and-tongued joints—hidden slot screwed joints—dowel joints—bridle joints—mortise-and-tenon joints—dovetail joints—mitered joints—framing joints—hinging and shutting joints—summary—review questions

CHAPTER 23

WOOD PATTERNMAKING...............................325

Tools—lumber—glue—shellac and varnish—fillets—patterns—cores—core prints—core boxes—draft—finish—shrinkage—blueprints—patternmaking joints—summary—review questions

CHAPTER 24

KITCHEN CABINET CONSTRUCTION 359

  Kitchen layout planning—kitchen wall cabinets—false cabinet
  walls—cabinet installation—unit kitchen cabinets—summary—review questions

INDEX 369

CHAPTER 1

Woods

Wood is our most versatile, most useful building material, and a general knowledge of the physical characteristics of various woods used in building is important for carpenter and handyman alike.

Wood may be classified:

1. Botanically. All trees that can be sawed into lumber or timbers belong to the division called Spermatophytes. This includes softwoods as well as hardwoods.
2. With respect to its density:
    a. Soft.
    b. Hard.
3. With respect to its leaves:
    a. Needle or scale leaved, botanically Gymnosperms, or conifers, commonly called softwoods. Most of them, but not all, are evergreens.
    b. Broad-leaved, botanically Angiosperms, commonly called hardwoods. Most are deciduous, shedding their leaves in the fall. Only one broad-leaved hardwood, the Chinese ginkgo, belongs to the subdivision Gymnosperms.
4. With respect to its shade or color:
    a. White or very light.
    b. Yellow or yellowish.
    c. Red.
    d. Brown.
    e. Black or nearly black.

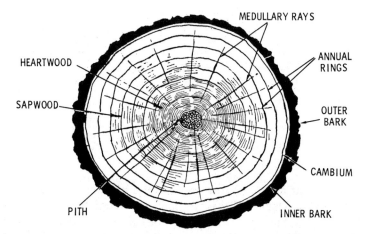

Fig. 1. Cross section of an oak nine years old, showing pitch, concentric rings comprising the woody part, the cambium layer, and the bark. The tree grows in concentric rings, or layers, with one layer added each year. The rings are also called annual rings. That portion of each ring formed in spring and early summer is softer, lighter colored, and weaker than that formed during the summer and is called spring wood. The denser, stronger wood formed later is called summer wood. The cells in the heartwood of some species are filled with various oils, tannins, and other substances, which make these timbers rot-resistant. There is practically no difference in the strength of heartwood and sapwood, if they weigh the same. In most species, only the sapwood can be readily impregnated with preservatives, a procedure used when the wood will be in contact with the ground.

    5. In terms of grain:
       a. Straight.
       b. Cross.
       c. Fine.
       d. Coarse.
       e. Interlocking.
    6. With respect to the nature of the surface when dressed:
       a. Plain—example, white pine.
       b. Grained—example, oak.
       c. Figured or marked—example, bird's-eye maple

As shown in Figs. 1 and 2, a tree consists of:

    1. Outer bark—living and growing only at the cambium layer. In most trees, the outside continually sloughs away.

2. In some trees, notably hickories and basswood, there are long tough fibers, called bast fibers, in the inner bark. In other trees, such as the beech, these bast fibers are absent.
3. Cambium layer. Sometimes this is only one cell thick. Only these cells are living and growing.
4. Medullary rays or wood fibers, which run radially.
5. Annual rings, or layers of wood.
6. Pith.

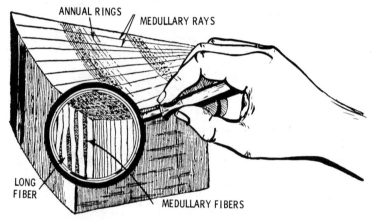

Fig. 2. A piece of wood magnified slightly to show its structure. The wood is made up of long, slender cells called fibers, which usually lie parallel to the pith. The length of these cells is often 100 times their diameter. Transversely, bands of other cells, elongated but much shorter, serve to carry sap and nutriments across the trunk radially. Also, in the hardwoods, long vessels or tubes, often several feet long, carry liquids up the tree. There are no sap-carrying vessels in the softwoods, but spaces between the cells may be filled with resins.

Around the pith, the wood substance is arranged in approximately concentric rings. The part nearest the pith is usually darker than the parts nearest the bark and is called the heartwood. The cells in the heartwood are dead. Nearer the bark is the sapwood, where the cells are living but not growing.

As winter approaches, all growth ceases, and each annual ring is separate and in most cases distinct. The leaves of deciduous trees, or trees that shed their leaves, and the leaves of some of the conifers, such as cypress and larch, fall, and the sap in the tree may freeze hard. The tree is dormant but not dead. With the warm days of the next spring, growth starts again strongly, and the cycle is repeated. The width of the annual rings varies greatly, from 30 to 40 or more per inch in some slow-growing species, to as few as 3 or 4 per inch in some of the quick-growing softwoods. The woods with the narrowest rings, because of the large percentage of summer wood, are generally strongest, although this is not always the case.

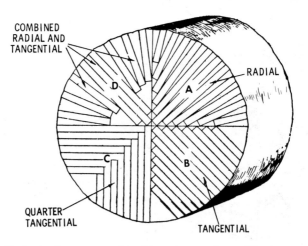

Fig. 3. *Methods of quartersawing. These are rarely used because waste is extensive.*

CUTTING AT THE MILL

When logs are taken to the mill, they may be cut in a variety of ways. One way of cutting is quartersawing. Fig. 3 shows three variations of this method. Here, each quarter is laid on the bark and ripped into quarters, as shown in the figure. Quartersawing is rarely done this way, though, because only a few wide boards are yielded; there is too much waste. More often, when quarter-sawed stock is required, the log is started as shown in Figs. 4 and 5, sawed until a good figure (pattern) shows, then turned over and sawed. This way there is little waste, and the boards are wide. In other words, most quartersawed lumber is resawed out of plain sawed stock.

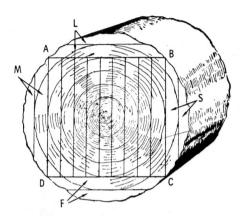

Fig. 4. Plain or bastard sawing, sometimes called flat or slash sawing. The log is first squared by removing boards MS and LF, giving the rectangular section ABCD. This is necessary to obtain a flat surface on the log.

The plain sawed stock, as shown, is simply flat sawing out of quartersawed. Quartersawed stock has its uses. Boards shrink most in a direction parallel with the annual rings, and door stiles and rails are often made of quartersawed material.

Lumber is sold by the board foot, meaning one 12-inch square of wood. Any stock under 2 inches thick is known as a board; over this and up to 4 inches it is known as lumber; and over 4 inches it is timber. The terms have become interchangeable, however, and are used interchangeably in this book.

Lumber, of course, is sold in nominal and actual size, the actual size being what the lumber is after being milled. As the years have gone by, the actual size has gotten smaller. A 2-by-4, for example, used to be an actual size of $1^9/_{16}$ inches by $3^9/_{16}$ inches. It is now 1½ inches by 3½ inches, and other boards go up or down in half-inch increments.

DEFECTS

The defects found in manufactured lumber, shown in Figs. 6 and 7, are grouped in several classes:

1. Those found in the natural log:
   a. Shakes.

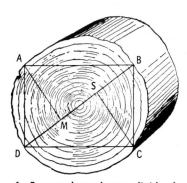

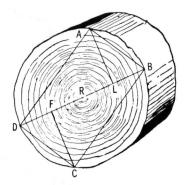

A. For one large beam, divide the longest diameter DB into three equal parts; erect perpendiculars at M and S, and join points thus obtained to form rectangle ABCD.

B. For one stiff beam, divide the longest diameter DB into four equal parts; erect perpendiculars at L and F, and join points thus obtained to form rectangle ABCD.

Fig. 5. Obtaining beams from a log.

    b. Knots.

    c. Pitchpockets.

  2. Those due to deterioration:

    a. Rot.

    b. Dote.

  3. Those due to imperfect manufacture:

    a. Imperfect machining.

    b. Wane.

    c. Machine burn.

    d. Checks and splits from imperfect drying.

Heart shakes, shown in Fig. 6, are radial cracks that are wider at the pith of the tree than at the outer end. This defect is more commonly found in old trees than in young vigorous saplings; it occurs frequently in hemlock.

A wind or cup shake is a crack following the line of the porous part of the annual rings and is curved by a separation of the annual rings. A wind shake may extend for a considerable distance up the trunk. Other explanations for wind shakes are expansion of the sapwood and wrenching from high wind (hence the name). Brown ash is especially susceptible to wind shake.

A star shake resembles a wind shake but differs from it in that the crack extends across the center of the trunk without any appearance of decay at that point; it is larger at the outside of the tree.

Dry rot, to which wood is so subject, is due to fungi; the name is misleading, as it only occurs in the presence of moisture and the absence of free air circulation.

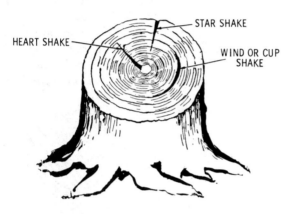

Fig. 6. Defects that can be found in lumber.

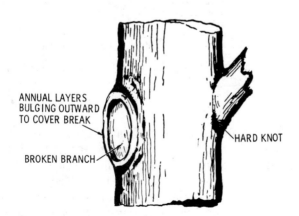

Fig. 7. Hard knot and broken branch showing nature's method of covering the break.

SELECTION OF LUMBER

A variety of factors must be considered when picking lumber for a particular project. For example, whether it is seasoned or not, that is, has it been dried naturally—the lumber is stacked up with air spaces between, as shown in Fig. 8—or artificially, as it is when dried in kilns. At any rate, the idea is to produce lumber with a minimum amount of moisture that will, therefore, warp least on the job. If your project requires nonwarping material, ask for kiln-dried lumber. If not, so-called green lumber will suffice. Green lumber is often used for framing (outside) where the slight warpage that occurs after it is nailed in place is not a problem. Green lumber is less costly than kiln-dried, of course.

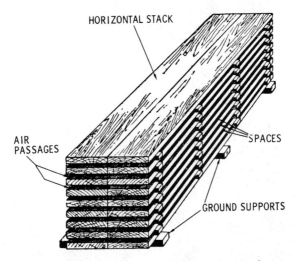

Fig. 8. Horizontal stack of lumber for air drying.

Another factor to consider is the grade of the lumber. The best lumber you can buy is Clear, which means the material is free of defects. Following this is Select, which has three subdivisions—Nos. 1, 2, and 3—with No. 1 the best of the Select lumber with only a few blemishes on one side of the board and few if any on the other. Last is Common. This is good wood, but it will have blemishes and knots that can interfere with a project if you want to finish it with a clear material.

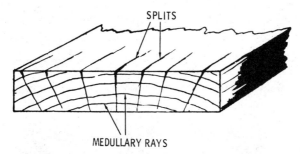

SPLITS

MEDULLARY RAYS

Fig. 9. A board with splits along the medullary rays. This condition is caused by too-rapid kiln drying.

Lumber has two grading systems, numerical and verbal. Numerically there are Nos. 1, 2, and 3, with No. 1 the best and No. 3 the least desirable. Roughly corresponding to these numbers are Construction, Standard, and Utility.

Wood may also be characterized as hardwood or softwood. These designations do not refer to the physical hardness of the wood, although hardwoods are normally physically harder than softwoods. The designation refers to the kinds of trees the wood comes from: Cone-bearing trees are softwoods, leaf-bearing trees are hardwoods. By far the most valuable softwood is pine, a readily available material in all sections of the country. (Of course the type of pine will depend on the particular area.) Hardwoods, which come in random lengths from 8 to 16 feet long and 4 to 16 inches wide and are all "Clear." Examples are mahogany, birch, oak, and maple. Hardwoods are usually much more expensive than softwoods.

In addition to the above, there are overall characteristics of the particular wood to consider. Following is a round-up of individual characteristics that will make the material more or less suitable for your project.

White or Gray Ash—Hard, heavy, and springy; light reddish brown heart; sapwood nearly white. Too hard to nail when dry.

Brown Ash—Not a framing timber, but an attractive trim wood. Brown heart, lighter sapwood. The trees often wind shake so badly that the heart is entirely loose. Attractive veneers are sliced from stumps and forks.

17

White Cedar, Northern—Light brown heart, sapwood thin and nearly white. Light, weak, soft, decay-resistant, holds paint well.

Western Red Cedar—Also called canoe cedar or shinglewood. Light, soft, straight-grained, small shrinkage, holds paint well. Heart is light brown, extremely rot-resistant. Sap quite narrow, nearly white. Used for shingles, siding, boat building.

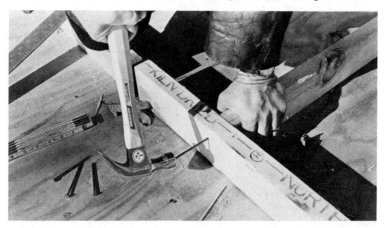

Fig. 10. Invaluable to the builder is 2 × 4 lumber. Note the classification "kiln dried" on one member. This stock will be used inside a structure. Courtesy of Vaughn & Bushnell.

Eastern Red Cedar, or Juniper—Pungent aromatic odor said to repel moths. Red or brown heartwood, extremely rot-resistant white sapwood. Used for lining clothes closets and chests and for fence posts.

Cypress—Probably our most durable wood for contact with the soil. Wood moderately light, close-grained, heartwood red to nearly yellow, sapwood nearly white. Does not hold paint well, but otherwise good for siding and outside trim. Its good looks also make it suitable for inside trim.

Gum, Red—Moderately heavy, interlocking grain; warps badly in seasoning; heart is reddish brown, sapwood nearly white. The sapwood may be graded out and sold as white gum, the heartwood as red gum, or together as unselected gum. Cuts into attractive veneers.

Hickory—A combination of hardness, weight, toughness, and strength found in no other native wood. A specialty wood, almost impossible to nail when dry. Not rot-resistant.

Hemlock, Eastern—Heartwood is pale brown to reddish, sapwood not distinguishable from heart. May be badly wind shaken. Brittle, moderately weak, not at all durable. Used for cheap, rough framing veneers.

Hemlock, Western—Heartwood and sapwood almost white with purplish tinge. Moderately strong, not durable, mostly used for pulpwood.

Locust, Black—Heavy, hard, strong; the heartwood is exceptionally durable. Not a framing timber. Used mostly for posts and poles.

Maple, Hard—Heavy, strong, hard, and close-grained; color light brown to yellowish. Used mostly for wear-resistant floors, and for furniture. Circularly growing fibers cause the attractive "birds-eye" grain in some trees. One species, the Oregon maple, occasionally contains the attractive "quilted" grain.

Maple, Soft—Softer and lighter than hard maple; lighter colored. Used for much the same purposes as hard maple, but not nearly so desirable.

Oak, White—Hard, heavy, tough, strong, and somewhat rot-resistant. Brownish heart, lighter sapwood. Desirable for trim and flooring, and one of our best hardwood framing timbers.

Oak, Red—Good framing timber, but not rot-resistant.

Pine, White, Western—Also called Idaho white pine. Creamy or light brown sapwood, sapwood thick and white. Used mostly for millwork and siding. Moderately light, moderately strong, easy to work, holds paint well.

Pine, Red or Norway—Resembles the lighter weight specimens of Southern yellow pine. Moderately strong and stiff, moderately soft, heartwood pale red to reddish brown. Used for millwork, siding, framing, and ladder rails.

Pine, Long-Leaf Southern Yellow—Heavy, hard, and strong, but not especially durable in contact with the soil. The sapwood takes preservative well. One of the most useful timbers for light framing.

Pine, Short-Leaf Southern Yellow—Quite satisfactory for light framing, and the sapwood is attractive as an interior finish.

Douglas Fir—A very plentiful commercial timber. Varies greatly in weight, color, and strength. Strong, moderately heavy, splintery, splits easily. Used in all kinds of construction; much is rotary cut for plywood.

Poplar, Yellow or Tulip—Our easiest-working native wood. Old growth has a yellow to brown heart. Sapwood and young trees are tough and white. Not a framing lumber, but used to a great extent for siding, where it may be marketed with cucumber magnolia, a botanical relative.

Redwood—One of our most durable and rot-resistant timbers. Light, soft, moderately high strength, heartwood reddish brown, sapwood white. Does not paint exceptionally well, as it often-times "bleeds" through. Used mostly for siding and outside trim.

Spruce, Sitka—Light, soft, medium strong, heart is light reddish brown, sapwood is nearly white, shading into the heartwood. Usually cut into boards, planing-mill stock, and boat lumber.

Spruce, Eastern—Stiff, strong, hard, and tough. Moderately lightweight, light color, little difference between heart and sapwood. Commercial Eastern spruce includes wood from three related species. Used for pulpwood, framing lumber, millwork, etc.

Spruce, Engelmann—Color, white with red tint. Straight grained, lightweight, low strength. Used for dimension lumber and boards, and for pulpwood. Extremely low rot resistance.

Tamarack, or Larch—Small to medium-sized trees; not much is sawed into framing lumber, but much is cut into boards. Yellowish brown heart, sapwood white. Much is cut into posts and poles.

Walnut, Black—Our most attractive cabinet wood. Heavy, hard, and strong, heartwood is a beautiful brown, sap nearly white. Mostly used for fine furniture, but some is used for fine interior trim. Somewhat rot-resistant. Used also for gunstocks.

Walnut, White or Butternut — Sapwood light to brown, heart light chestnut brown with an attractive sheen. The cut is small, mostly going into cabinet work and interior trim. Moderately light, rather weak, not rot-resistant.

Decay of Lumber

Decay of lumber is the result of one cause, and one cause only, the work of certain low-order plants called fungi. All of these organisms require water and air to live, grow, and multiply; consequently, wood that is kept dry, or that is dried quickly after wetting, will not decay. Similarly, wood that is kept submerged in water will soften, but it will not decay, for the air supply is shut off, and timber set deep in the ground, such as piling, which is shut off from the air, is practically permanent.

There is no such thing as "dry rot"; however, the term is rather loosely used sometimes when speaking of about any type of rot or any dry and decayed wood. Although such rotten wood may be dry when observed, it was wet while decay was progressing. This kind of decay is often found inside living, growing trees, but it occurs only in the presence of water.

OTHER MATERIALS

Plywood

In addition to boards and lumber, there are other materials that carpenters and handymen have come to rely on. Among the most important is plywood.

The most familiar type of plywood used in the United States is made from Douglas fir. Short logs are chucked into a lathe, and a thin, continuous layer of wood is peeled off. This thin layer is straightened, cut to convenient sizes, covered with glue, laid up with the grain in successive plies crossing, and subjected to heat and pressure. This is the plywood of commerce, one of our most useful building materials.

All plywood has an odd number of plies, allowing the face plies to have parallel grain while the lay-up is "balanced" on each side of a center ply. This process equalizes stresses set up when the board dries or when it is subsequently wetted and dried.

Plywood is also graded, and the grading system has changed over the years. Most important, however, there is Exterior or Interior plywood—which refers to the type of glue used to bond plies—and both come in various grades. The face of the plywood has a letter grade—A, B, C, or D. Grade A means the face has no defects—it's perfect. Grade B means there are some defects; perhaps an area has a small patch. Grade C allows checks (splits) and small knotholes. Grade D allows large knotholes.

In theory you should be able to get a wide variety of grades and types in the lumberyard, but in reality you commonly find, at this writing, AC Interior, AD Interior, AC Exterior, and CD, which is sheathing.

Fig. 11. For building something fast and not having to worry about warpage, you can't beat plywood. Courtesy of The American Plywood Assn.

In selecting plywood the rule is simple: Just pick what is right for the job at hand. If one side is going to be hidden, for example, you do not need a high grade.

22

Particle Board

In the manufacture of particleboard the structure of the wood is not broken down, but simply reduced to flakes, or particles, which are bound together with a synthetic resin, often ureaformaldehyde or phenol-formaldehyde. The boards are then cured under heat and moderate pressure.

Particleboard is much less costly than plywood, but it is rough on saw blades and difficult to nail at the edges. Indeed, it cannot be nailed at the edges. For saving money, though, it cannot be beaten. Like plywood, particleboard is available in 4 × 8-foot sheets of various thicknesses.

Hardboard

Hardboard is made from wood pulp, and it usually contains no binder other than the slightly thermoplastic lignin in the wood. The board is formed under heat and heavy pressure. One side of the board is smooth; the other may or may not have a textured surface. It may be tempered by impregnating the board with oil and then baking it.

Hardboard is good for light jobs and for lining floors prior to installing flooring.

SUMMARY

There are many basic methods of preparing lumber for the market. It is necessary that it be seasoned (moisture removed), this process being classed as natural and artificial. Natural seasoning consists of exposing sawed lumber to freely circulating air. Artificial drying, or kiln drying, is accomplished in most cases by forcing heated air over the lumber.

The selection of various grades of lumber to be used for any specific purpose is often left to the discretion of the carpenter. Defects often found in manufactured lumber are shakes, knots, pitch-pockets, imperfect machining, machine burns, and splits from imperfect drying. Knots, coarse grain, and other defects may or may not reduce the strength of the lumber, depending on their location in the piece.

REVIEW QUESTIONS

1. How may lumber be seasoned? Explain.
2. What are some of the defects found in lumber?
3. What should a person look for when purchasing framing lumber?

CHAPTER 2

Nails

Up to the end of the Colonial period, all nails used in the United States were handmade. They were forged on an anvil from nail rods, which were sold in bundles. These nail rods were prepared either by rolling iron into small bars of the required thickness or by the much more common practice of cutting plate iron into strips by means of rolling shears.

Just before the Revolutionary War, the making of nails from these rods was a household industry among New England farmers. The struggle of the Colonies for independence intensified an inventive search for shortcuts to mass production of material entering directly or indirectly into the prosecution of the war; thus came about the innovation of cut nails made by machinery. With its coming, the household industry of nail making rapidly declined. At the close of the 18th century, 23 patents for nail-making machines had been granted in the United States, and their use had been generally introduced into England, where they were received with enthusiasm.

In France, lightweight nails for carpenter's use were made of wire as early as the days of Napoleon I, but these nails were made by hand with a hammer. The handmade nail was pinched in a vise with a portion projecting. A few blows of a hammer flattened one end into a head. The head was beaten into a countersunk depression in the vise, thus regulating its size and shape. In the United States, wire nails were first made in 1851 or 1852 by William Hersel of New York.

In 1875, Father Goebel, a Catholic priest, arrived from Germany and settled in Covington, Kentucky; there he began the manufacture of wire nails that he had learned in his native land. In 1876, the American Wire and Screw Nail Company was formed under Father Goebel's leadership. As the production and consumption of wire nails increased, the vogue of cut nails, which dominated the market until 1886, declined.

The approved process in the earlier days of the cut-nail industry was as follows: Iron bars, rolled from hematite or magnetic pig, were fagotted, reheated to a white heat, drawn, rolled into sheets of the required width and thickness, and then allowed to cool. The sheet was then cut across its length (its width being usually about a foot) into strips a little wider than the length of the required nail. These plates, heated by being set on their edge on hot coals, were seized in a clamp and fed to the machine, end first. The cut-out pieces, slightly tapering, were squeezed and headed up by the machine before going to the trough.

The manufacture of tacks, frequently combined with that of nails, is a distinct branch of the nail industry, affording much room for specialties. Originally it was also a household industry, and was carried on in New England well into the 18th century. The wire, pointed on a small anvil, was placed in a pedal-operated vise, which clutched it between jaws furnished with a gauge to regulate the length. A certain portion was left projecting; this portion was beaten with a hammer into a flat head.

Antique pieces of furniture are frequently held together with iron nails that are driven in and countersunk, thus holding quite firmly. These old-time nails were made of four-square wrought iron and tapered somewhat like a brad but with a head which, when driven in, held with great firmness.

The raw material of the modern wire nail factory is drawn wire, just as it comes from the wire-drawing block. The stock is low-carbon Bessemer or basic open-hearth steel. The wire, feeding from a loose reel, passes between straightening rolls into the gripping dies, where it is gripped a short distance from its end, and the nailhead is formed by an upsetting blow from a heading tool. As the header withdraws, the gripping dies loosen, and the straightener carriage pushes the wire forward by an amount equal to the length of the nail. The cutting dies advance from the sides of the frame and clip off the nail, at the same time forming its characteristic chisel point. The gripping dies have already seized the wire again, and an ejector flips the nail out of the way just as the header comes forward and heads the next nail. All these motions are induced by cams and eccentrics on the main shaft of the machine, and the speed of production is at a rate of 150 to 500 or more complete cycles per minute. At this stage, the nails are covered with a film of drawing lubricant and oil from the nail machine, and their points are frequently adorned with whiskers—a name applied to the small diamond-shaped pieces stamped out when the point is formed and which are occasionally found on the finished nail by the customer.

These oily nails, in lots of 500 to 5000 pounds, are shaken with sawdust in tumbling barrels from which they emerge bright and clean and free of their whiskers, ready for weighing, packing, and shipping.

THE "PENNY" SYSTEM

This method of designating nails originated in England. Two explanations are offered as to how this interesting designation came about. One is that the six penny, four penny, ten penny, etc., nails derived their names from the fact that one hundred nails cost six pence, four pence, etc. The other explanation, which is the more probable of the two, is that one thousand ten-penny nails, for instance, weighed ten pounds. The ancient, as well as the modern, abbreviation for penny is *d*, being the first letter of the Roman coin denarius; the same abbreviation in early history was used for the English pound in weight. The word *penny* has persisted as a term in the nail industry.

NAIL CHARACTERISTICS

Nails are the carpenter's most useful fastener, and a great variety of types and sizes are available to meet the demands of the industry. One manufacturer claims to produce more than 10,000 types and sizes. Some common types are illustrated in Fig. 1.

Nails also have a variety of characteristics, that is, different points, shanks, finishes, and material. The following shapes of points are available:

NAILS

1. Common blunt pyramidal.
2. Long sharp.
3. Chisel-shaped.
4. Blunt, or shooker.
5. Side-sloped.
6. Duck-bill, or clincher.

The heads may be:

1. Flat.
2. Oval or oval countersunk.
3. Round.
4. Double-headed.

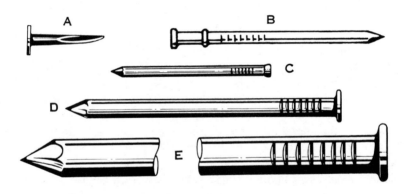

Fig. 1. Various nails grouped as to general size: A, tack; B, sprig or dowel pin; C, brad; D, nail; E, spike.

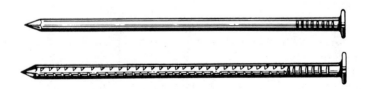

Fig. 2. Smooth and barbed box nails, lbd size (shown full size). Note the sharp point and thin, flat head.

Each of the features or characteristics makes the nail better suited for the job at hand. For example, galvanized nails are weather-resistant, double-headed nails are good for framing where they can be installed temporarily with the second head exposed for easy pulling, barbed nails are good when extra holding power is required.

Tacks

Tacks are small, sharp-pointed nails that usually have tapering sides and a thin, flat head. The regular lengths of tacks range from ⅛ to 1⅛ inches. The regular sizes are designated in ounces, according to Table 1. Tacks are usually used to secure carpet or fabric.

Table 1. Wire Tacks

Size oz.	Length in.	No. per pound	Size oz.	Length in.	No. per pound	Size oz.	Length in.	No. per pound
1	1/8	16,000	4	7/16	4000	14	13/16	1143
1 1/2	3/16	10,666	6	9/16	2666	16	7/8	1000
2	1/4	8000	8	5/8	2000	18	15/16	888
2 1/2	5/16	6400	10	11/16	1600	20	1	800
3	3/8	5333	12	3/4	1333	22	1 1/16	727
....	....	....	....	....	....	24	1 1/8	666

Brads

Brads are small slender nails with small deep heads; sometimes, instead of having a head, they have a projection on one side. There are several varieties adapted to many different requirements. Brad sizes start at about ½ inch and end at 1½ inches; beyond this size they are called finishing nails.

Nails

The term "nails" is popularly applied to all kinds of nails except extreme sizes, such as tacks, brads, and spikes. Broadly speaking, however, it includes all of these. The most generally used are called common nails, and are regularly made in sizes from 1 inch (2d) to 6 inch (60d), as shown in Table 2 and Figs. 4, 5, and 6. Some special types of nails are illustrated in Figs. 6 through 10.

Fig. 3. Brads are the smallest nails you can get. They are used to attach molding and the like. Courtesy of The American Plywood Assn.

Spikes

One can think of a spike as an extra large nail, sometimes quite a bit larger. Generally, spikes range from 3 to 12 inches long and are thicker than common nails. Point style varies, but a spike is normally straight for ordinary uses, such as securing a gutter. However, a spike can also be curved or serrated, or cleft to make extracting or drawing it out very difficult. Spikes in larger sizes are used to secure rails to ties, in the building of docks, and for other large-scale projects.

If you have a very large job to do, it is well to know the holding power of nails. In most instances, this information will not be required, but in more than a few cases it is.

Tests on nails (and spikes) ranging in size from $6d$ to $60d$ are shown in Table 4. It is interesting to note, in view of the rela-

Table 2. Common Nails

	Plain			Coated			
Size	Length in.	Gauge No.	No. per pound	Length in.	Gauge No.	No. per Keg	Net Wgt. pounds
2d	1	15	876	1	16	85,700	79
3d	1¼	14	568	1⅛	15½	54,300	64
4d	1½	12½	316	1⅜	14	29,800	61
5d	1¾	12½	271	1⅝	13½	25,500	70
6d	2	11½	181	1⅞	13	17,900	65
7d	2¼	11½	161	2⅛	12½	15,300	72
8d	2½	10¼	106	2⅜	11½	10,100	71
9d	2¾	10¼	96	2⅝	11½	8900	68
10d	3	9	69	2⅞	11	6600	63
12d	3¼	9	63	3⅛	10	6200	80
16d	3½	8	49	3¼	9	4900	80
20d	4	6	31	3¾	7	3100	83
30d	4½	5	24	4¼	6	2400	84
40d	5	4	18	4¾	5	1800	82
50d	5½	3	14	5¼	4	1300	79
60d	6	2	11	5¾	3	1100	82

tively small force required to withdraw nails, that spikes take tremendous pulling power. In one test it was found that a spike ⅜ inch in diameter driven 3½ inches into seasoned yellow pine required 2000 pounds of force for extraction. And the denser the material, the more difficult the extraction is. The same spike required 4000 pounds of force to be withdrawn from oak and 6000 pounds from well-seasoned locust.

Roofing Nails

A specialized but often-used nail is the roofing nail. This has a barbed shank and a large head, which makes it good at holding down shingles and roofing paper felt without damage—the material cannot pull readily through the head.

Such nails come in a variety of sizes but usually ⅜ inch to 1¼ inch long with the nail sized to the material thickness.

Drywall Nails

As the name implies, these are for fastening drywall—Sheetrock. The shank of the nail is partially barbed and the head countersunk so that if the nail bites into the stud, it takes a good bite. Drywall nails come in a variety of lengths for use with different thicknesses of Sheetrock.

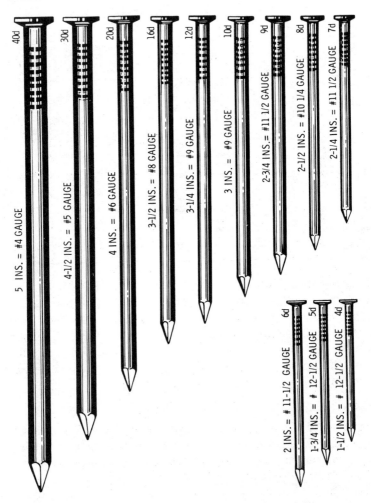

Fig. 4. Common wire nails—the standard nail for general use is regularly made in sizes from 1 inch (2d) to 6 inches (60d).

Masonry Nails

Masonry nails are cut, that is stamped, out of a sheet of metal rather than drawn and cut the way wire nails are. A masonry nail is made of very hard steel and is case-hardened. It has a variety of uses but the most common is probably for securing

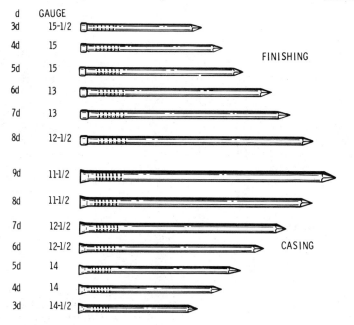

d	GAUGE	
3d	15-1/2	
4d	15	
5d	15	FINISHING
6d	13	
7d	13	
8d	12-1/2	
9d	11-1/2	
8d	11-1/2	
7d	12-1/2	
6d	12-1/2	CASING
5d	14	
4d	14	
3d	14-1/2	

Fig. 5. Casing and finishing nails (shown full size). Note the difference in the head shape and size. The finishing nail is larger than a casing nail of equal length, but a casing nail is stronger.

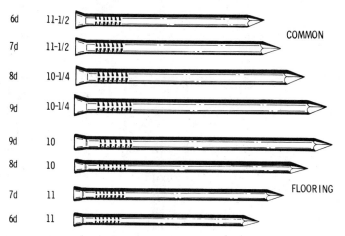

Fig. 6. Flooring and common nails (shown full size). Note the variation in head shape and gauge number.

NAILS

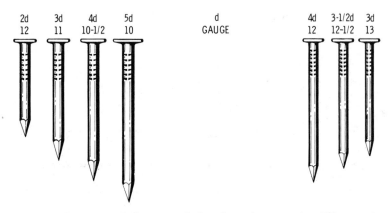

Fig. 7. A few sizes of slating and shingle nails. Note the difference in wire gauge.

Fig. 8. Hook-head, metal-lath nail. This is a bright, smooth nail with a long, thin flat head, made for application of metal lath. It is also made blued or galvanized.

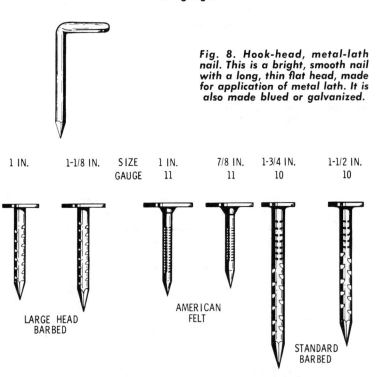

Fig. 9. Various roofing nails (shown full size).

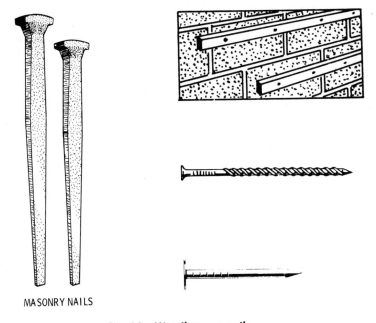

MASONRY NAILS

Fig. 10. Miscellaneous nails.

Table 3. Withdrawal Force
(lbs. per sq. in. of surface)

Wood	Wire Nail	Cut Nail
White Pine	167	405
Yellow Pine	318	662
White Oak	940	1216
Chestnut		683
Laurel	651	1200

Table 4. Holding Power of Nails and Spikes (Withdrawal)

Size of spikes	5×¼ in. sq.	6×¼	6×½	5×⅜
Length driven in	4¼ in.	5 in.	5 in.	4¼ in.
Pounds resistance to drawing, average lbs.	857	857	1691	1202
max. lbs.	1159	923	2129	1556
From 6 to 9 tests each min. lbs.	766	766	1120	687

studs or furring to block walls. Safety is important when doing any kind of nailing, but it is particularly important when using masonry nails to wear protective goggles to guard the eyes against flying chips.

Panel Nails

When installing paneling, the idea is to hide the nail, and that is what panel nails are for. They come in a variety of colors so that when driven into the paneling, their heads don't show.

Spiral Nails

The most tenacious of all nails in terms of holding power is the spiral nail, also known as the drive screw. Its shank is spiral so that as the nail is driven, it turns and grips the wood. Its main use is to secure flooring, but it is also useful on rough carpentry.

Corrugated Fasteners

You are probably familiar with the corrugated fastener, although you may not know its technical name. At any rate, the fastener is a small section of corrugated metal with one sharpened and one flat edge. Corrugated fasteners are often used for making boxes or joining wood sections edge to edge. The finish is normally blued, and they come in a variety of sizes.

Staples

Many varieties of staples are available, from ones used to secure cable and fencing to posts (such staples are always galvanized) to ones used in the various staple guns. The former generally range in size from ⅞ inch to 1¼ inches, and some are designed—the so-called slash point—so the legs spread when the staple is driven in place; this makes it grip better. Other types of staples are covered in Chapter 17 on Fastening Tools.

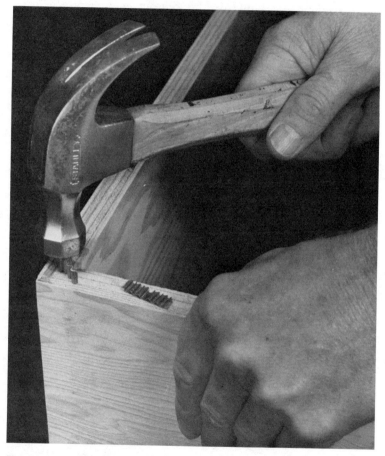

Fig. 11. Corrugated nails do not have great holding power, but they are handy. Courtesy of The American Plywood Assn.

Framing Fasteners

Framing fasteners fall into a category all their own, and they are very useful when you have any framing to do. A few companies make them. Basically, they are stamped (16 or 18 gauge) metal pieces with predrilled nail (or screw) holes. You set the fastener between the pieces and drive the nails through the holes to lock the members together. The result is a very strong connection. It should be noted that not all building codes accept them, so their use should be checked out beforehand.

Figs. 12A and B illustrate some typical framing fasteners.

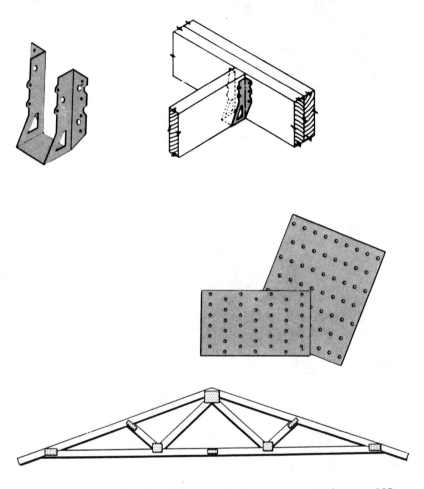

Fig. 12A. One useful kind of framing fastener is the joist hanger. 12B. Perforated plates such as shown can be used to make trusses. Courtesy of Teco.

SELECTING NAIL SIZE

In selecting nails for jobs, size is crucial. First consideration is the diameter. Short, thick nails work loose quickly. Long, thin nails are apt to break at the joints of the lumber. The simple rule to follow is to use as long and as thin a nail as will drive easily.

Definite rules have been formulated by which to determine the size of nail to be used in proportion to the thickness of the board that is to be nailed:

1. When using box nails in lumber of medium hardness, the penny of the nail should not be greater than the thickness, in eighths of an inch, of the board into which the nail is being driven.
2. In very soft woods, the nails may be one penny larger, or even in some cases, two pennies larger.
3. In hard woods, nails should be one penny smaller.
4. When nailing boards together, the nail point should penetrate within ¼ inch of the far side of the second board.

The kind of wood is, of course, a big factor in determining the size of nail to use: The dry weight of the wood is the best basis for the determination of its grain substance or strength. The greater its dry weight, the greater its power to hold nails. However, the splitting tendency of hard wood tends to offset its additional holding power. Smaller nails can be used in hard lumber than in soft lumber, as shown in Fig. 13. Positive rules governing the size of nails to be used as related to the density of the wood cannot be laid down. Experience is the best guide.

DRIVING NAILS

In most cases it is not necessary to drill pilot holes for nails to avoid splitting the wood. However, in some instances it is advisable to first drill holes nearly the size of the nail before driving, to guard against it (Fig. 14). Also, in fine work, where a large number of nails must be driven, such as in boat building, holes should be driven. This step prevents crushing the wood and possible splitting because of the large number of nails driven

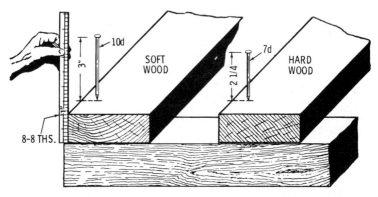

Fig. 13. Application of rules 2 and 3 in determining the proper size of nail to use.

Fig. 14. To prevent a nail from splitting wood, a pilot hole is sometimes drilled. Pilot-hole drilling is common when using screws. Courtesy of The American Plywood Assn.

through each plank. The size of drill for a given size nail should be found by experiment.

The right and wrong ways to drive nails are shown in Fig. 15. Fig. 16 illustrates the necessity of using a good hammer to drive a nail. The force that drives the nail is due to the inertia of the hammer. This inertia depends on the suddenness with which its motion is brought to rest on striking the nail. With hardened steel, there is practically no give, and all the energy possessed by the hammer is transferred to the nail. On a hammer made

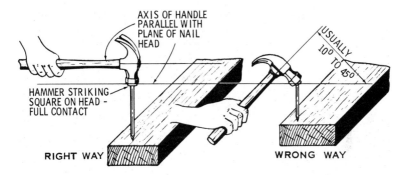

Fig. 15. Right and wrong ways to drive a nail. Hit the nail squarely on the head. The handle should be horizontal when the hammer head hits a vertical nail.

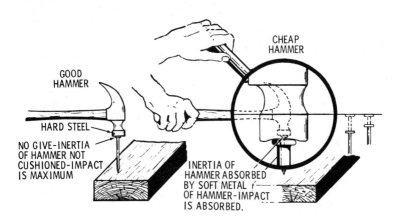

Fig. 16. Why a cheap hammer should not be used.

with soft and/or inferior metal, all the energy is not transferred to the nail; therefore, the power per blow is less than with hardened steel.

SUMMARY

Nails are the carpenter's most useful fastener, and many nail types and sizes are available to meet the demands of the indus-

Table 5. Approximate Number

American Steel & Wire Co's. Steel Wire Gauge	Length									
	3/16	1/4	3/8	1/2	5/8	3/4	7/8	1	1 1/8	1 1/4
3/8	....	....	....	....	....	....	....	29	26	23
5/16	....	....	....	....	....	....	....	43	38	34
1	....	....	....	....	....	....	....	47	44	40
2	....	....	....	....	....	....	....	60	54	48
3	....	....	....	....	....	....	....	67	60	55
4	....	....	....	....	....	....	....	81	74	66
5	....	....	....	....	....	....	....	90	81	74
6	....	....	....	213	174	149	128	113	101	91
7	....	....	....	250	205	174	148	132	120	110
8	....	....	....	272	238	198	174	153	139	126
9	....	....	....	348	286	238	213	185	170	152
10	....	....	....	469	373	320	277	242	616	196
11	....	....	....	510	117	366	323	285	254	233
12	....	....	....	740	603	511	442	397	351	327
13	....	....	1356	1017	802	688	590	508	458	412
14	....	2293	1664	1290	1037	863	765	667	586	536
15	....	2899	2213	1619	1316	1132	971	869	787	694
16	....	3932	2770	2142	1708	1414	1229	1099	973	872
17	....	5316	3890	2700	2306	1904	1581	1409	1253	1139
18	....	7520	5072	3824	3130	2608	2248	1976	1760	1590
19	....	9920	6860	5075	4132	3508	2816	2556	2284	2096
20	18620	14050	9432	7164	5686	4795	4230	3596	3225	2893
21	23260	17252	12000	8920	7232	6052	5272	4576	4020	3640
22	28528	21508	14676	11776	9276	7672	....	....	....	....
23	35864	27039	18026	13519	10815	9013	....	....	....	....
24	44936	34018	22678	17008	13607	11339	....	....	....	....
25	57357	43243	28828	21622	17297	14414	....	....	....	....

try. On any kind of construction work, an important consideration is the type and size of nails to use.

An important factor in selecting nails is size. Long, thin nails will break at the joints of the lumber. Short, thick nails will work loose quickly. The kind of wood is, of course, a big factor in determining the size of nail to use.

of Wire Nails per Pound

American Steel & Wire Co's. Steel Wire Gauge	Length										
	1½	1¾	2	2¼	2½	2¾	3	3½	4	4½	5
⅜	20	17	15	15	12	11	11	8.9	7.9	7.1	6.4
5⁄16	29	25	22	20	18	16	15	13	11	10	9.0
1	34	29	26	23	21	20	18	16	14	12	11
2	41	35	31	28	25	23	21	18	16	14	13
3	47	41	36	32	29	27	25	21	18	16	15
4	55	48	41	37	34	31	29	25	22	20	18
5	61	52	45	41	38	35	32	28	24	22	21
6	76	65	58	52	47	43	39	34	29	26	24
7	92	78	70	61	55	51	47	40	35	31	28
8	106	93	82	74	66	61	56	48	42	38	34
9	128	112	99	87	79	71	67	58	50	45	41
10	165	142	124	111	100	91	84	71	62	55	49
11	200	171	149	136	122	111	103	87	77	69	61
12	268	229	204	182	161	149	137	118	103	95	87
13	348	297	260	232	209	190	175	153	138	123	110
14	459	398	350	312	278	256	233	201	176	157	140
15	578	501	437	390	351	317	290	256	220	196	177
16	739	635	553	496	452	410	370	318	277	248	226
17	956	831	746	666	590	532	486	418	360	322	295
18	1338	1150	996	890	820	740	680	585	507	448	412
19	1772	1590	1390	1205	1060	970	895	800	....	....	....
20	2412	2070	1810	1620	1450	1315	1215	1035	....	....	....
21	3040	2665	2310	2020	1830	....	....	....	....	....	....
22	....	....	....	....	....	....	....	....	....	....	....
23	....	....	....	....	....	....	....	....	....	....	....
24	....	....	....	....	....	....	....	....	....	....	....
25	....	....	....	....	....	....	....	....	....	....	....

These approximate numbers are an average only, and the figures given may be varied either way, by changes in the dimensions of the heads or points. Brads and on-head nails will run more to the pound than table shows, and large or thick-headed nails will run less.

REVIEW QUESTIONS

1. What is nail holding power?
2. Explain the "penny" nail system.
3. What should be considered when selecting a nail for a particular job?
4. Name and describe five kinds of useful nails.

CHAPTER 3

Screws

Wood screws have a couple of advantages over nails. First, screws are harder to pull out. Pull on a screw and pull on a nail—the screw will give greater resistance. Second, should you tire of an item at some future date, screws usually let you disassemble it without great travail. It is possible to damage the work if it is nailed together and you want to take it apart.

Screws are normally used to fasten things—hinges, knobs, etc.—to structures and in the assembly of a variety of woodworking projects. They are not used in heavy building simply because in this type of work things are built so that there is a minimum of stress on the fasteners and the withdrawal resistance is not required. Indeed, if stress were created, even the most tenacious screw could not stand up much better than a nail—which is to say very little.

The wood screw consists of a gimlet point, a threaded portion, and a shank and head, which may be straight slot or Phillips. More about these later.

Screws of many types are made for specialized purposes, but stock wood screws are usually obtainable in either steel or brass, and, more rarely, are made of high-strength bronze. Three types of heads are standard: the flat countersunk head, with the included angle of the sloping sides standardized at 82°; the round head, whose height is also standardized, but whose contour seems to vary slightly among the products of different manufacturers; and the oval head, which combines the contours of the flat head and the round head. All of these screws are available with the Phillips slot, or crossed slots, instead of the usual single straight slot.

The Phillips slot allows a much greater driving force to be exerted without damaging the head, and it is more sightly than the usual straight-slotted head. By far the greater part of all wood screws used, probably 75% or more, are of the flat-head type.

MATERIAL

For ordinary purposes, steel screws, with or without protective coatings, are commonly used. In boat building or other such work where corrosion will probably be a problem if screws are used, the screws should be of the same metal or at least the same *type* of metal as the parts they contact. While it is possible and indeed probable that a single brass screw driven through an aluminum plate, if it is kept dry, will show no signs of corrosion, many brass screws driven through the aluminum plate in the presence of water or dampness will almost certainly show signs, perhaps serious signs, of galvanic corrosion.

DIMENSIONS OF SCREWS

When ordering screws, you must know what constitutes length. It varies with head type. The overall length of a 2-inch flat-head screw is not the same as a 2-inch round-head screw. As detailed in Fig. 1, the flat-head length is from the top of the head to the tip of the screw, while the round-head screw measures from under the head and the oval-head screw measures from the edge of the top.

It should also be noted that, unlike ordinary wire gauges, the 0 in the screw gauge, shown in Fig. 2, indicates the diameter of the smallest screw, and the diameter of the screws *increases* with the number of the gauge.

SHAPE OF THE HEAD

You can find a variety of head shapes on screws, but the three standard shapes are flat, round, and oval. These usually will more than suffice.

All of these heads are available in either the straight-slotted or Phillips type.

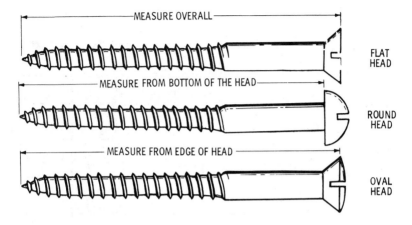

Fig. 1. Various wood screws and how their length is measured.

The other forms may be regarded as special or semispecial, that is, carried by large dealers only or obtainable only on special order.

Flat heads are necessary in some cases, such as on door hinges, where any projection would interfere with the proper working of the hinge; flat-head screws are also employed on finish work where flush surfaces are desirable. The round and oval heads are normally ornamental, left exposed.

The diameter of the head in relation to the gauge number of the screw is shown in Table 1. Some of the many special screw heads available are shown in Fig. 3.

HOW TO DRIVE A WOOD SCREW

Driving wood screws is made easier by drilling a pilot hole in the wood. Indeed, this may be the only way to do it. Consult Table 2 for the size drill to use in drilling a shank-clearance hole. This hole should be slightly smaller than the shank diameter of the screw and about ¾ the shank length for soft and medium-hard wood. For very hard wood, the hole should be shank length.

Again consult Table 2 to determine the size of drill to use in drilling the pilot hole. This hole should be equal in diameter to

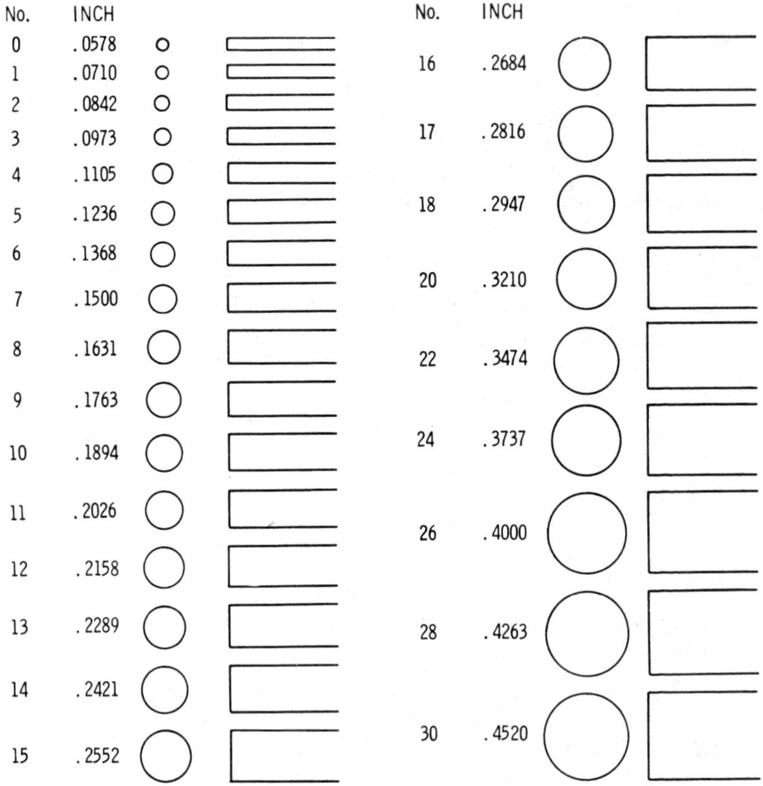

No.	INCH				No.	INCH		
0	.0578							
1	.0710				16	.2684		
2	.0842							
3	.0973				17	.2816		
4	.1105							
5	.1236				18	.2947		
6	.1368							
7	.1500				20	.3210		
8	.1631				22	.3474		
9	.1763							
10	.1894				24	.3737		
11	.2026				26	.4000		
12	.2158							
13	.2289				28	.4263		
14	.2421							
15	.2552				30	.4520		

Fig. 2. Wood screw gauge numbers.

the root diameter of the screw thread and about ¾ the thread length for soft and medium-hard woods. For extremely hard woods, the pilot-hole depth should equal the thread length.

If the screw being inserted is the flat-head type, the hole should be countersunk. A typical countersink is shown in Fig. 5.

The foregoing process involves three separate steps. All of these can be performed at once by using a device of the type shown in Fig. 6. This tool will drill the pilot hole, the shank-clearance hole, and the countersink all in one operation. Stanley calls its device the Screw-Mate. The Stanley company also makes a Screw-Sink, which counterbores—you can set the head

Table 1. Head Diameters

Screw Gauge	Screw Diameter	Head Diameter		
		Flat	Round	Oval
0	0.060	0.112	0.106	0.112
1	0.073	0.138	0.130	0.138
2	0.086	0.164	0.154	0.164
3	0.099	0.190	0.178	0.190
4	0.112	0.216	0.202	0.216
5	0.125	0.242	0.228	0.242
6	0.138	0.268	0.250	0.268
7	0.151	0.294	0.274	0.294
8	0.164	0.320	0.298	0.320
9	0.177	0.346	0.322	0.346
10	0.190	0.371	0.346	0.371
11	0.203	0.398	0.370	0.398
12	0.216	0.424	0.395	0.424
13	0.229	0.450	0.414	0.450
14	0.242	0.476	0.443	0.476
15	0.255	0.502	0.467	0.502
16	0.268	0.528	0.491	0.528
17	0.282	0.554	0.515	0.554
18	0.394	0.580	0.524	0.580
20	0.321	0.636	0.569	0.636
22	0.347	0.689	0.611	0.689
24	0.374	0.742	0.652	0.742
26	0.400	0.795	0.694	0.795
28	0.426	0.847	0.735	0.847
30	0.453	0.900	0.777	0.900

of the screw beneath the surface, then plug the hole with a wood plug cut with a plug cutter from matching wood.

These tools are made in many sizes, one for each screw size, and they are available in complete sets or separately. The screw size is marked on the tool.

STRENGTH OF WOOD SCREWS

Table 2 gives the safe resistance, or safe load (against pulling out), in pounds per linear inch of wood screws when inserted across the grain. For screws inserted with the grain, use 60% of these values.

The lateral load at right angles to the screw is much greater than that of nails. For conservative designing, assume a safe resistance of a No. 20 gauge screw at double that given for nails of the same length, when the full length of the screw thread penetrates the supporting piece of the two connected pieces.

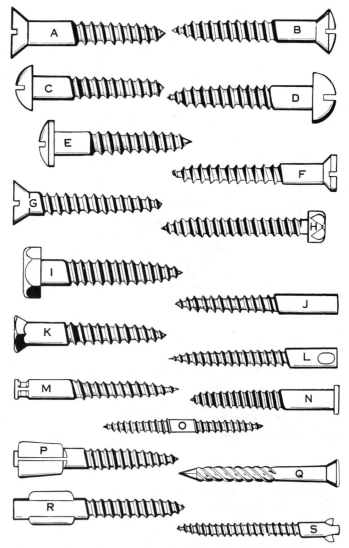

Fig. 3. Various wood screws showing the variety of head shapes available. A, flat head; B, oval head; C, round head; D, piano head; E, oval fillister head; F, countersunk fillister head; G, felloe; H, close head; I, hexagon head; J, headless; K, square bung head; L, grooved; M, pinched head; N, round bung head; O, dowel; P, winged; Q, drive; R, winged; S, winged head. Heads A through G may be obtained with Phillips-type head. Most will never be needed.

Table 2. Safe Loads for Wood Screws

Kind of Wood	Gauge Number							
	4	8	12	16	20	24	28	30
White oak	80	100	130	150	170	180	190	200
Yellow pine	70	90	120	140	150	160	180	190
White pine	50	70	90	100	120	140	150	160

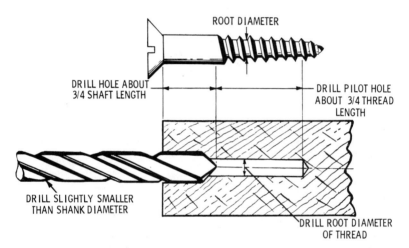

Fig. 4. Drilling shank-clearance and pilot holes.

Fig. 5. A typical countersink.

LAG SCREWS

By definition, a lag screw, shown in Fig. 7, is a heavy-duty wood screw provided with a square or hexagonal head so that it

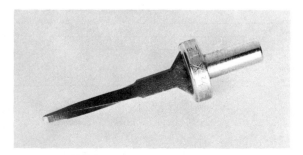

Fig. 6. A tool for drilling pilot hole, shank-clearance hole, and counter-sink in one operation.

Fig. 7. Ordinary lag screw.

may be turned by a wrench. Lag screws are large, heavy screws used where great strength is required, such as for heavy timber work. Table 3 gives the dimensions of ordinary lag screws.

How To Put in Lag Screws

First, bore a hole slightly larger than the diameter of the shank to a depth that is equal to the length that the shank will penetrate (see Fig. 8). Then bore a second hole at the bottom of the first hole equal to the root diameter of the threaded shank and to a depth of approximately one-half the length of the threaded portion. The exact size of this hole and its depth will, of course,

Table 3. Lag Screws

Length	3	3½	4	4½	5	5½	6	6½	7	7½	8	9	10	11	12
Dia.	5/16 to 7/8	5/16 to 1	5/16 to 1	5/16 to 1	5/16 to 1	5/16 to 1	5/16 to 1	7/16 to 1	7/16 to 1	7/16 to 1	7/16 to 1	7/16 to 1	1/2 to 1	1/2 to 1	1/2 to 1

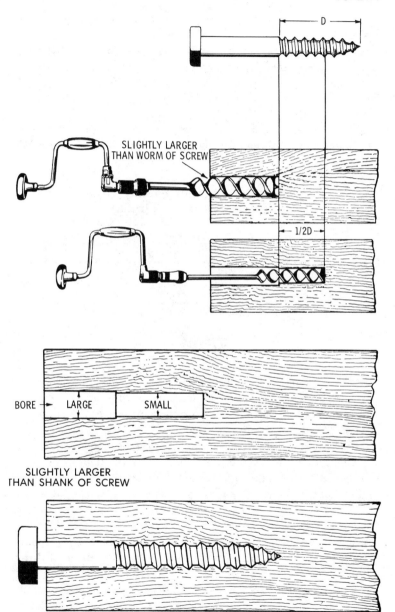

Fig. 8. Drilling holes for lag screws.

depend on the kind of wood; the harder the wood, the larger the hole.

The resistance of a lag screw to turning is enormous when the hole is a little small, but this can be considerably decreased by smearing the threaded portion of the screw with beeswax.

Strength of Lag Screws

Table 4 gives the safe resistance, to pull out load, in pounds per linear inch of thread for lag screws when inserted across the grain.

Table 4. Safe Loads for Lag Screws
(Inserted across the grain)

Kind of Wood	Diameter of Screw in Inches				
	1/2	5/8	3/4	7/8	1
White pine	590	620	730	790	900
Douglas fir	310	330	390	450	570
Yellow pine	310	330	390	450	570

SUMMARY

Wood screws are often used in carpentry because of their advantage over nails in strength. They are used in installing various types of building hardware because of their great resistance to pulling out and because they are more or less readily removed in case of repairs or alterations.

There are generally three standard types of screw heads—the flat countersunk head, the round head, and the oval head, all of which can be obtained in crossed slot, single straight slot, or Phillips slot. For ordinary purposes, steel screws are commonly used, and in wood-screw applications, probably 75% or more are flat-head screws.

Lag screws are heavy-duty wood screws that are provided with a square or hexagonal head so that they are installed with a wrench. These are large, heavy screws that are used where great strength is needed, such as for heavy timber and beam installations. Holes are generally bored into the wood since the diameter of lag screws is large.

Table 5. Standard Wood Screw Proportions

Screw Numbers	A	B	C	D	Number of Threads per Inch
0	....	....	....	0.0578	30
1	....	....	....	0.0710	28
2	0.1631	0.0454	0.030	0.0841	26
3	0.1894	0.0530	0.032	0.0973	24
4	0.2158	0.0605	0.034	0.1105	22
5	0.2421	0.0681	0.036	0.1236	20
6	0.2684	0.0757	0.039	0.1368	18
7	0.2947	0.0832	0.041	0.1500	17
8	0.3210	0.0809	0.043	0.1631	15
9	0.3474	0.0984	0.045	0.1763	14
10	0.3737	0.1059	0.048	0.1894	13
11	0.4000	0.1134	0.050	0.2026	12.5
12	0.4263	0.1210	0.052	0.2158	12
13	0.4427	0.1286	0.055	0.2289	11
14	0.4790	0.1362	0.057	0.2421	10
15	0.5053	0.1437	0.059	0.2552	9.5
16	0.5316	0.1513	0.061	0.2684	0
17	0.5579	0.1589	0.064	0.2815	8.5
18	0.5842	0.1665	0.066	0.2947	8
20	0.6368	0.1816	0.070	0.3210	7.5
22	0.6895	0.1967	0.075	0.3474	7.5
24	0.7421	0.2118	0.079	0.3737	7
26	0.7421	0.1967	0.084	0.4000	6.5
28	0.7948	0.2118	0.088	0.4263	6.5
30	0.8474	0.2270	0.093	0.4546	6
....	....	....	....	....	....

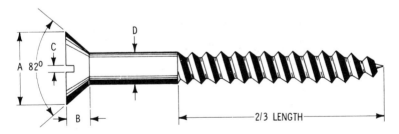

REVIEW QUESTIONS

1. Name the three basic head shapes of wood screws.
2. What are the three screw slots used on wood screws?
3. What type of wood screw is used where great strength is required?
4. What type of head is used on lag screws? Why?
5. What is meant by the root diameter of a screw?

CHAPTER 4

Bolts

Although many bolts are available, the classic or standard bolt is designed for assembling things where strength is required. A nut is normally part of the device.

KINDS OF BOLTS

One commonly used bolt is the carriage bolt, which got its name from its prime early use: assembling horsedrawn carriages.

To install a carriage bolt, a hole is bored that is the diameter of the shank. The bolt is then slipped into the hole, and a hammer is used to pound it down so that the neck on it seats well in the hole. A nut on the other end completes the job, something that can be done without having to hold the other end of the bolt.

Another bolt is the machine bolt. It is used on metal and wood items. The machine bolt is slipped into the hole and a wrench used to hold its large square head on one end while another wrench is used to tighten a nut in place. Excellent tension is possible.

The stove bolt is slotted like a screw and is used for assembling metal items. The stove bolt has lag screw threads on one end and machine screw threads on the other with a smooth portion between. One prime use for the stove bolt is in assembling furniture. The lag screw portion goes into the frame while the other end screws into a socket in the leg. For carpenters and do-it-yourselfers, a stud bolt is useful for hanging things from large beams.

A. Square head and square nut.

B. Hexagon head and hexagon nut.
Fig. 1. Machine bolts.

Fig. 1. Machine bolts.

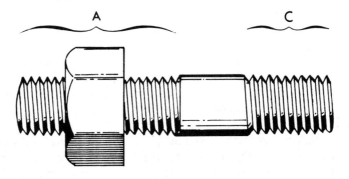

Fig. 2. Stud bolt with hexagon nut. A is the nut end and C is the attachment end.

Eyebolts have a looped end and are useful for hanging clothesline.

Various bolts are illustrated in Fig. 3.

MANUFACTURE OF BOLTS

The bolt-and-nut industry in America was started on a small scale in Marion, Connecticut, in 1818. In that year, Micah Rugg,

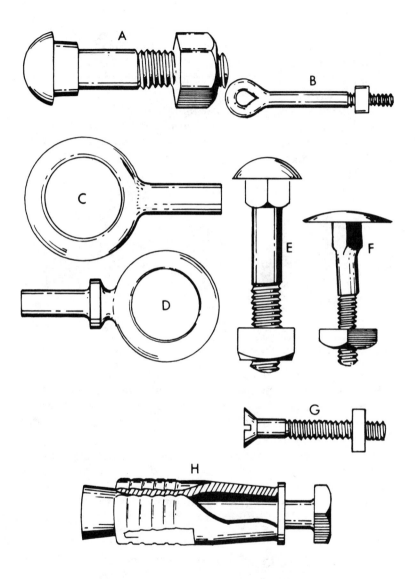

Fig. 3. Various bolts. In the figure, A is a railroad track bolt; B, a welded eye bolt; C, a plain forged eye bolt; D, a shouldered eye bolt; E, a carriage bolt; F, a step bolt; G, a stove bolt; H, an expansion bolt.

59

a country blacksmith, made bolts by the forging process. The first machine used for this purpose was a device known as a heading block, which was operated by a foot treadle and a connecting lever. The connecting lever held the blank while it was being driven down into the impression in the heading block by a hammer. The square iron from which the bolt was made was first rounded so that it could get into the block.

At first, Rugg only made bolts to order, and charged at the rate of 16 cents apiece. This industry developed quite slowly until 1839 when Rugg went into partnership with Martin Barnes. Together they built the first exclusive bolt-and-nut factory in the United States at Marion, Connecticut.

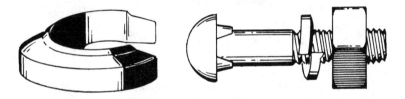

Fig. 4. A lock washer. When the nut is screwed onto the bolt, it strikes the rib on the washer, which is much harder than the nut. The rib on the washer is forced into the nut, thus preventing the nut from loosening.

Bolts were first manufactured in England in 1838 by Thomas Oliver of Darlston, Staffordshire. His machine was built on a somewhat different plan from that of Rugg's, but no doubt was a further development of the first machine. Oliver's machine was known as the "English Oliver."

The construction of the early machines was carefully kept secret. It is related that in 1842, a Mr. Clark had his bolt-forging machine located in a room separated from the furnaces by a thick wall. The machine received the heated bars through a small hole cut in the wall; the forge man was not even permitted to enter the room.

PROPORTIONS AND STRENGTH OF BOLTS

Ordinary bolts are manufactured in certain "stock sizes." Table 1 gives these sizes for bolts from ¼ inch up to 1¼ inches, with the length of thread.

Table 1. Properties of U.S. Standard Bolts
(U.S. Standard or National Coarse Threads)

Diameter	Number of Threads per inch (National Coarse Thread)	Head	Head	Head
1/4	20	3/8	13/32	1/2
5/16	18	1/2	35/64	43/64
3/8	16	9/16	5/8	3/4
7/16	14	5/8	11/16	53/64
1/2	13	3/4	53/64	1
9/16	12	7/8	31/32	1 5/32
5/8	11	15/16	1 1/32	1 1/4
3/4	10	1 1/8	1 15/64	1 1/2
7/8	9	1 5/16	1 29/64	1 47/64
1	8	1 1/2	1 21/32	1 63/64
1 1/8	7	1 11/16	1 55/64	2 15/64
1 1/4	7	1 7/8	2 1/16	2 31/64
1 3/8	6	2 1/16	2 17/64	2 47/64
1 1/2	6	2 1/4	2 31/64	2 63/64
1 5/8	5 1/2	2 7/16	2 11/16	3 15/64
1 3/4	5	2 5/8	2 57/64	3 31/64
1 7/8	5	2 13/16	3 3/32	3 47/64
2	4 1/2	3	3 5/16	3 63/64

Table 2. National Fine Threads

Diameter	Threads per inch
1/4	28
5/16	24
3/8	24
7/16	20
1/2	20
9/16	18
5/8	18
3/4	16
7/8	14
1	14

For many years, the coarse-thread bolt was the only type available. Now, bolts with a much finer thread, called the National Fine thread, have become easily available. These have hex heads and hex nuts. They are finished much better than the stock coarse-thread bolts and consequently are more expensive. Cheap rolled-thread bolts, with the threaded portions slightly upset, should not be used by the carpenter. When they are driven into a hole, either the hole is too large for the body of the bolt or the threaded portion reams it out too large for a snug fit. Good bolts have cut threads that have a maximum diameter no larger than the body of the bolt.

When a bolt is to be selected for a specific application, Table 3 should be consulted.

Table 3.   Proportions and Strength of U.S. Standard Bolts

Bolt Diameter	Area at Bottom of Threads	Tensile Strength		
		10,000 lbs/in^2	12,500 lbs/in^2	17,500 lbs/in^2
1/4	0.027	270	340	470
5/16	0.045	450	570	790
3/8	0.068	680	850	1190
7/16	0.093	930	1170	1630
1/2	0.126	1260	1570	2200
9/16	0.162	1620	2030	2840
5/8	0.202	2020	2520	3530
3/4	0.302	3020	3770	5290
7/8	0.419	4190	5240	7340
1	0.551	5510	6890	9640
1 1/8	0.693	6930	8660	12,130
1 1/4	0.890	8890	11,120	15,570
1 3/8	1.054	10,540	13,180	18,450
1 1/2	1.294	12,940	16,170	22,640
1 5/8	1.515	15,150	18,940	26,510
1 3/4	1.745	17,450	21,800	30,520
1 7/8	2.049	20,490	25,610	35,860
2	2.300	23,000	28,750	40,250

Example—How much of a load may be applied to a 1-inch bolt for a tensile strength of 10,000 pounds per square inch?

Referring to Table 3, we find on the line of a 1-inch bolt a value of 5510 pounds corresponding to a stress on the bolt of 10,000 pounds per square inch.

Example—What size bolt is required to support a load of 4000 pounds for a stress of 10,000 pounds per square inch?

area at root of thread = given load ÷ 10,000
= 4000 ÷ 10,000 = 0.400 sq. in.

Referring to Table 3, in the column headed "Area at Bottom of Thread," we find 0.419 square inch to be the nearest area; this corresponds to a ⅞-inch bolt.

Of course, for the several given values of pounds stress per square inch, the result could be found directly from the table, but the calculation above illustrates the method that would be employed for other stresses per square inch not given in the table.

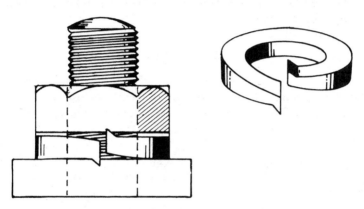

Fig. 5. Positive lock washer. The barbs force themselves deeply into the nut and the backing.

Example—A butt joint with fish plates is fastened by six bolts through each member. What size bolts should be used, allowing a shearing stress of 5000 pounds per square inch in the bolts, when the joint is subjected to a tensile load of 20,000 pounds?

load carried per bolt = 20,000 ÷ number of bolts
= 20,000 ÷ 6 = 3333 lbs.

Each bolt is in double shear, hence:

equivalent single shear load = ½ of 3333 = 1667 lbs.

and

$$area\ per\ bolt = \frac{1667}{5000} = 0.333\ sq.\ in.$$

Referring to Table 3, the nearest area is 0.302, which corresponds to a ¾-inch bolt. In the case of a dead, or "quiescent," load, ¾-inch bolts would be ample; however, for a live load, take the next larger size, or ⅞-inch bolts.

The example does not give the size of the members, but the assumption is they are large enough to safely carry the load. In practice, all parts should be calculated as described in the chapter on the strength of timbers. The ideal joint is one so proportioned that the total shearing stress of the bolts equals the tensile strength of the timbers.

OTHER BOLTS

A few other bolts are very useful for the carpenter and handyman, although these bolts are unlike standard bolts. They are the toggle bolt; the hollow wall anchor commonly known by one name, the Molly Bolt; and the expansion shield or bolt.

The toggle bolt is used to hang things on hollow wall. A hole is drilled in the wall. Spring-actuated springs on the bolt are compressed, and the bolt is slipped through the hole. Inside the wall the bolt springs open. When the bolt (actually a machine screw) is tightened, the item is snugged up against the wall.

A Molly Bolt is like a toggle bolt but its expansive wings are built on and it can be reused. To use it, the screw is withdrawn from the socket part of the device, which is left in the wall; the screw is slipped through the item to be hung and rescrewed into the socket.

Another useful device is the expansion bolt. (This bolt is illustrated in Fig. 3H.) It works like a Molly Bolt except that it is designed for use in masonry.

SUMMARY

A bolt is generally regarded as a rod having a head at one end and a threaded portion on the other to receive a nut. The nut is usually considered as forming a part of the bolt. Bolts are used to connect two or more pieces of material when a very strong connection is required.

Various forms of bolts are manufactured to meet the demands

and requirements of the building trade. The common machine bolt has a square or hexagonal head. The carriage bolt has a round head; the stove bolt has a round or countersunk head with a single slot. Lock washers are used to prevent nuts from loosening. Other useful bolts are the toggle, Molly, and expansion.

REVIEW QUESTIONS

1. What type of head is generally found on a machine bolt?
2. What is meant by "threads per inch"?
3. Explain the purpose of lock washers.
4. What is an expansion bolt?
5. What is tensile load?

CHAPTER 5

The Workbench

There is hardly a shop for any purpose that does not require a workbench, especially the carpenter's shop. Many of these shops, unfortunately, have only a makeshift bench made of waste stock, crudely put together, with no part of it strong enough and no place to attach a vise. Such benches really can be more of a hindrance than a help.

Substantial benches are manufactured and sold for all purposes, but you can easily construct the type of bench best suited to your requirements.

Fig. 1 shows a quickly constructed bench such as would be used at a construction site. A bench such as that shown in Fig. 2 should be constructed for permanent installation in a shop.

In the last few years foreign manufacturers have introduced a number of compact, well-made workbenches to America, and one manufacturer, Black and Decker, has come up with a real innovation, called the Workmate. This is virtually a table, but it can be adjusted to hold all kinds of items and can be carried to the job.

The height of the bench should be dictated by the character of the work to be done—high for light work and low for heavy work. The height of the person who is to use the bench should also be considered. In general, carpenter's benches are made 33 inches high, while those for cabinetmakers and patternmakers are from 2 to 4 inches higher.

WORKBENCH ATTACHMENTS

A number of devices are used with workbenches to make the jobs to be done easier. Useful accessories are:

1. Vises.
2. Support pegs.

3. Bench stop.
4. Bench hook.

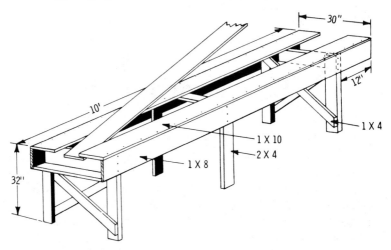

Fig. 1. A quickly constructed workbench for temporary use. Lumber required: five 2'' × 4'' × 8'; one 1'' × 4'' × 16'; two 1'' × 8'' × 10'; and
three 1'' × 10'' × 10'. The center legs add rigidity, but on such a short
bench, they are not absolutely necessary.

Vises

The general construction of a woodworker's vise is shown in
Fig. 4. There are numerous types in use. Usually there is a large,
or main, vise of iron construction at one end of the table (as
shown in Fig. 4) and frequently a smaller or supplemental iron
vise at the other end for small work.

For patternmaking work, a quick-closing vise is frequently
used. It is advisable to face any iron vise with a wood or leather
covering to prevent marking or denting the lumber, especially
when soft woods are used.

Support Pegs

The function of the main bench vise is to prevent the wood
from moving while being worked, as with a plane or chisel. In
these operations, the wood receives a pressure that tends to rotate it in the plane of the vise jaws, the latter acting as a pivot.

In the case of a long board, this turning force, or torque,

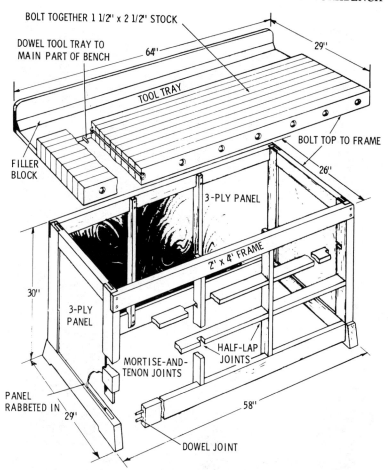

BOLT TOGETHER 1 1/2" x 2 1/2" STOCK

DOWEL TOOL TRAY TO
MAIN PART OF BENCH — 64"

29"

TOOL TRAY

BOLT TOP TO FRAME

FILLER
BLOCK

26"

3-PLY PANEL

2" X 4' FRAME

30"

3-PLY
PANEL

HALF-LAP
JOINTS

MORTISE-AND-
TENON JOINTS

PANEL
RABBETED IN
29"

58"

DOWEL JOINT

Fig. 2. Construction details of a workbench designed for appearance as well as convenience. The 2'' × 4'' rails, corner posts, and base members provide a sturdy bench that will give many years of satisfactory service.

would become extremely great when a downward pressure is applied at the far end of the board, thereby requiring the vise to be screwed up rather firmly to prevent turning. To avoid this, the bench is provided with supporting pegs, which carry the weight of the board and prevent it from turning. A vertical row of holes for the pegs should be provided at the middle and at the right end of the bench.

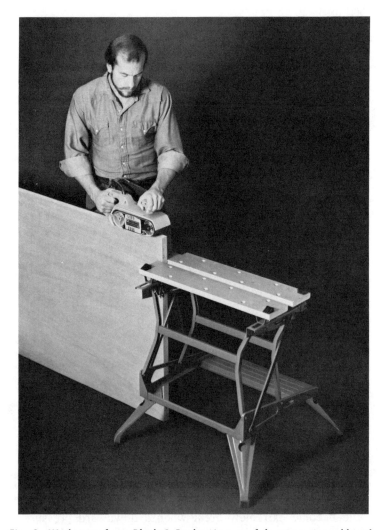

Fig. 3. Workmate from Black & Decker is a useful, compact workbench.
Courtesy of Black & Decker.

Bench Stop

This accessory is intended to prevent any lengthwise movement of the work while it is being tooled; that is, it prevents

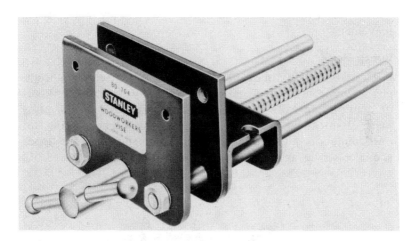

Fig. 4. A woodworker's vise with guides to keep the jaws in parallel. The inner jaw is fastened to the bench and supports a fixed nut in which the screw rotates. The screw moves the outer jaw which has attached to it two rods that slide through guide sleeves to keep the jaws parallel at all times.

endwise movement of a board while the board is being planed. As usually constructed for this purpose, a bench stop (Fig. 5) consists of a metal casing that is designed to set in flush with the bench and has a horizontal toothed plate that works in vertical guides. A screw adjustment is provided so that the plate may be set flush with the top of the table (when not in use) or a little

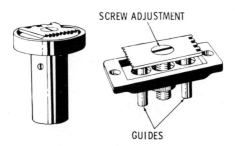

SCREW ADJUSTMENT

GUIDES

Fig. 5. Round and rectangular forms of a bench stop. These adjust by a center screw from flush to as high as required for the work. The round bench stop is fitted by boring a hole the diameter of the stop with an expansion bit and a deeper center with the proper size of bit. The rectangular bench stop is shallow and should be mortised in flush with the bench top.

71

above so as to engage the end of the work and prevent endwise movement.

Bench Hook

This is virtually a movable stop that can be used at right angles to the front of the bench. It serves many purposes for holding and putting work together. When it is desired to saw off a piece of stock, the bench hook is placed on the bench (Fig. 6); one shoulder is set against the edge of the bench, and the upper shoulder serves as a stop for the work while sawing.

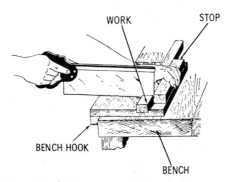

Fig. 6. *Typical bench hook and method of use when sawing to size with backsaw. The bench hook is used for a variety of operations, such as odd sawing and chiseling, and also serves to prevent the workbench from being marred by such operations.*

Storing Tools

Different craftsmen have different ways of storing their tools. One of the best for quick access to the tools is a panel made of Pegboard that can be mounted on the wall behind the workbench. The Pegboard has holes in it to accept hangers, which are inserted; tools are then hung from them. Prior to mounting the panel, the Pegboard should be given a coat or two of primer followed by a finish coat of enamel in a color that goes well in the area.

SUMMARY

There is hardly a workshop—especially a woodworking shop—that does not require a workbench of some kind. Sub-

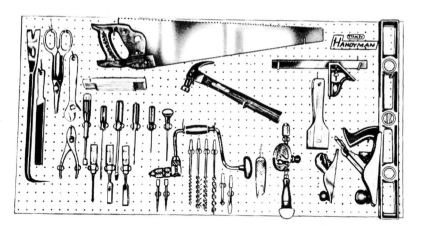

Fig. 7. A tool panel showing arrangement of tools.

stantial benches are manufactured and sold for all purposes, but a carpenter or the do-it-yourselfer can construct the type of bench best suited to his or her requirements.

A bench should be constructed with care as a permanent installation in the shop. The height and width of the bench should be regulated by the character of the work to be done—high enough for light work and low enough for heavy work. The height of the person who is going to use the bench should also be considered.

The general construction of a bench should be heavy enough, or at least anchored to the floor, in order to withstand hard, rough work without moving. Various bench attachments or devices, such as vises, bench stops, and bench hooks, are used with workbenches to ease the operations to be performed.

REVIEW QUESTIONS

1. What size lumber should be used when making a good sturdy workbench?
2. What are workbench attachments? Name a few.
3. Why should a workbench be heavy or at least anchored to the floor?
4. Why should you always have a tool panel?

CHAPTER 6

Carpenters' Tools

While it may seem obvious, the obvious should sometimes be stated: The more tools a carpenter has, the easier the job will usually be.

There is, of course, another idea to live by, as it were: Buy good-quality tools. Poor-quality tools make jobs much more difficult—and even unsafe. The handle may break off a tool, or an edge may fail just when you need it most.

One good way of telling quality is to compare and contrast tools of varying prices. It will soon become obvious which tools are well made, even if you aren't an expert on quality. Good tools have a certain heft, a feel, a finish that is undeniable.

The tools you'll want to own can be grouped in a number of categories. These are, of course, hand tools. In succeeding volumes, power tools will be suggested.

1. Guiding and testing tools.
   Straightedge
   Square
      try square
      miter square
      combined try and miter square
      framing square
      combination square
   Sliding "T" bevel
   Miter box
   Level

    Plumb bob
    Plumb rule

2. Marking tools.
    Chalk line
    Carpenter's pencil
    Ordinary pencil
    Scratch awl
    Scriber
    Compass and dividers

3. Measuring tools.
    Tape measure
    Various folding rules
    Rules with attachments
    Marking gauges

4. Holding tools.
    Horses or trestles
    Clamps
    Vises

5. Toothed cutting tools.
    Saws
        hand
        circular
        band
    Files and rasps
    Sandpaper

6. Sharp-edged cutting tools.
    Chisels
        paring
        firmer
        corner
        gouge
        tang and socket
        butt pocket and mill
    Drawknife

7. Axes and hatchets.
    Hatchet
    Axe

8. Smooth facing tools.
   Spokeshave
   Planes
     jack
     fore
     jointer
     smoothing
     block

9. Boring tools.
   Awl
   Gimlets
   Augers
   Drills
   Hollow augers
   Countersinks
   Reamers

10. Fastening tools.
    Hammers
    Screwdrivers
    Wrenches

11. Sharpening tools.
    Abrasives
    Grinding wheels (electric powered)
    Oilstones
      natural
      artificial

SUMMARY

A good set of tools is a necessity, and it is important to buy only the best, regardless of cost. A careful selection from standard brands, examining them to be sure there are no visible defects, is always a good practice.

The type and number of hand tools required depend on individual preferences and the work to be done, although a good basic list will consist of about thirty tools. While the list of hand tools may be increased to a very large number, most ordinary woodworking projects can be performed with a strictly limited number of tools.

REVIEW QUESTIONS

1. What tools are considered measuring tools?
2. Under what category are chalk lines and scratch awls?
3. Name the various types of wood planes.
4. What are tooth cutting tools?
5. What are holding tools?

CHAPTER 7

Guiding and Testing Tools

Before a board is cut or a nail is driven home, guiding and testing tools come into play. Happily, a variety of these tools are available, and they make the carpenter capable of the precision that many jobs demand. Following is a useful roundup.

STRAIGHTEDGE

This tool is used to guide a pencil or scriber when marking a straight line and when testing a faced surface, such as the edge of a board, to determine if it is straight. Anything having an edge known to be straight, such as the edge of a steel square, may be used; however, a regular straightedge is preferable.

The straightedge may be made either of wood or steel, and its length may be from a few inches to several feet. For ordinary work, a carpenter can make a sufficiently accurate straightedge from a strip of good straight-grained wood, as shown in Fig. 1, but for precision work, a steel straightedge, such as the three shown in Fig. 2, should be used. Wood is objectionable for precision work because of its tendency to warp or spring out of shape.

Fig. 3 shows the correct and incorrect methods of holding a straightedge as a guiding tool, and Fig. 4 shows how and how not to hold the pencil when marking stock.

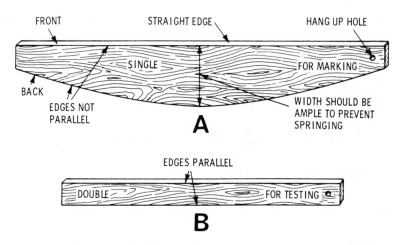

Fig. 1. Wooden straightedges. When well made, they are sufficiently accurate for ordinary use; A, single straightedge; B, double straightedge.

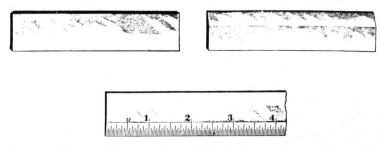

Fig. 2. Typical steel straightedges. These tools are used where straight lines are to be scribed or where surfaces must be tested for flatness. Depending on their use, straightedges are made in lengths of from 12 to 72 inches, widths of 1⅜ to 2½ inches, and thicknesses of 1/16 to ½ inch.

SQUARE

This tool is a 90°, or right-angle, standard and is used for marking or testing work. There are several common types of squares, as shown in Fig. 5; they are:

1. Try square.
2. Miter square.
3. Combined-try-and-miter square.

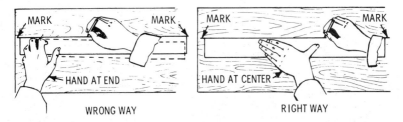

Fig. 3. *The incorrect and correct methods of using a straightedge as a guiding tool. To properly secure the straightedge, the hand should press firmly on the tool at its center, with the thumb and other fingers stretched wide apart.*

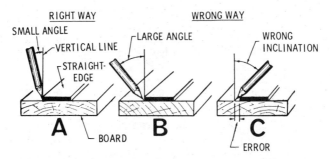

Fig. 4. *Right and wrong inclinations of the pencil in marking with the straightedge. The pencil should not be inclined from the vertical more than is necessary to bring the pencil lead in contact with the guiding surface of the straightedge (A). When the pencil is inclined more, and pressed firmly, considerable pressure is brought against the straightedge, tending to push it out of position (B). If the inclination is in the opposite direction, the lead recedes from the guiding surface, thus introducing an error which is magnified when a wooden straightedge is used because of the greater thickness of the straightedge (C).*

4. Framing or so-called "steel" square.
5. Combination square.

Try Square

In England, this is called the trying square, but here it is simply the try square. It is so called probably because of its frequent use as a testing tool when squaring up mill-planed stock. The ordinary try square used by carpenters consists of a steel blade set at right angles to the inside face of the stock in which it is held. The stock is made of some type of hardwood and is

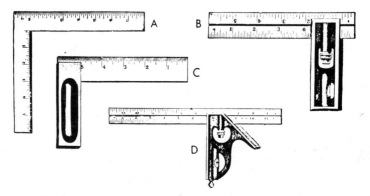

Fig. 5. Various types of squares. In the illustration, A represents a steel square; B, a double try square; C, a try square; D, a combination square. This last square consists of a graduated steel rule with an accurately machined head. The two edges of the head provide for measurements of 45° and 90°.

faced with brass in order to preserve the wood from damage.

The usual sizes of try squares have blades ranging 3 to 15 inches long. The stock is approximately ½ inch thick, with the blade inserted midway between the sides of the stock. The stock is made thicker than the blade so that its face may be applied to the edge of the wood and the steel blade may be laid on the surface to be marked. Usually the blade is provided with a scale of inches divided into eighths.

Miter and Combined Try-and-Miter Squares

The term "miter," strictly speaking, signifies any angle except a right angle, but, as applied to squares, it means an angle of 45°.

In the miter square, the blade (as in the try square) is permanently set but at an angle of 45° with the stock, as shown in Fig. 6.

A try square may be made into a combined try-and-miter square when the end of the stock to which the blade is fastened is faced off at 45°, as along the line *MS* in Fig. 7. When the 45° face (*MS*) of the stock is placed against the edge of a board, the blade will be at an angle of 45° with the edge of the board, as in Fig. 8.

An improved form of the combined try-and-miter square is

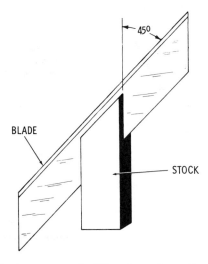

Fig. 6. A typical miter square; it differs from the ordinary try square in that the blade is set at an angle of 45° with the stock, and the stock is attached to the blade midway between its ends.

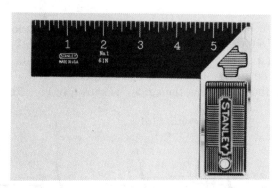

Fig. 7. A combined try square and miter square. Because of its short 45° face (MS), it is not as accurate as the miter square, but it answers the purpose for ordinary marking and the necessity for extra tools.

shown in Fig. 9. Because of the longer face (*LF*) as compared with the short face (*MS*) in Fig. 7, the blade describes an angle of 45° with greater precision. Its worst disadvantage is that it is awkward to carry because of its irregular shape. However, its precision greatly outweighs its disadvantage.

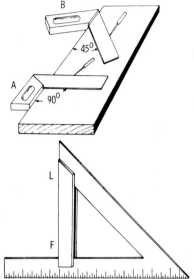

Fig. 8. The combined try-miter square as used for a 90° marking at A and a 45° marking at B.

Fig. 9. An improved form of the combined try-miter square.

A square having a blade not exactly at the intended angle is said to be out of true, or simply "out," and good work cannot be done with a square in this condition. A square should be tested, and if found to be out, should be returned.

The method of testing the square is shown in Fig. 10. This test

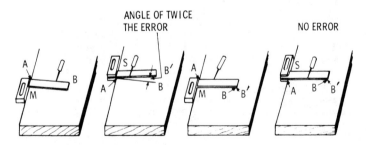

Fig. 10. The method of testing a try square. If the square is "out" (angle at 90°), scribed lines AB and AB' for positions M and S of the square (left side) will not coincide. Angle BAB' is twice the angle of error. If the square is perfect, lines AB and AB' for positions M and S will coincide (right side).

should be made not only at the time of purchase but frequently afterward, because the tool may become imperfect from a fall or rough handling.

Under no circumstances should initials or other markings be stamped on the brass face of the ordinary try square, because the burrs which project from bending the brass face will throw the square out of truth; for this reason, manufacturers will not take back a square with any marks stamped on the brass face.

Framing or "Steel" Square

While the framing square is commonly called the steel square, it is true that all types of squares may be obtained that are made entirely of steel. It is properly called a framing square because with its framing table and various other scales, it is adapted especially for use in house framing, although its range of usefulness makes it valuable to any woodworker. The tool and some of its functions are shown in Fig. 11.

The framing square consists of two essential parts—the tongue and the body, or blade. The tongue is the shorter, narrower part, and the body is the longer, wider part. The point at which the tongue and the body meet on the outside edge is called the heel.

There are several grades of squares, including polished, nickeled, blued, and royal copper. The blued square with figures and scales in white is perhaps the most desirable. A size that is widely used has an 18-inch body and a 12-inch tongue, but there are many uses which require the largest size whose body measures 24 by 2 inches and whose tongue measures 16 or 18 by 1½ inches.

The feature that makes this square so valuable a tool is its numerous scales and tables. These are:

1. Rafter or framing table.
2. Essex table.
3. Brace table.
4. Octagon scale.
5. Hundredths scale.
6. Inch scale.
7. Diagonal scale.

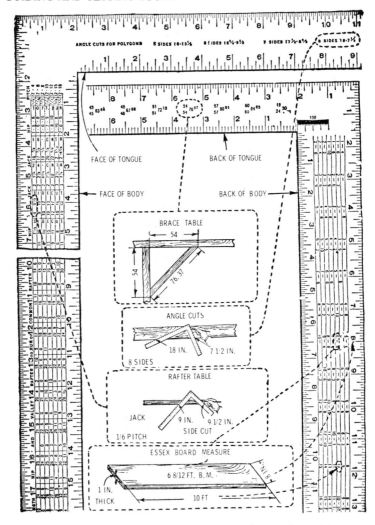

ANGLE CUTS FOR POLYGONS   6 SIDES 18-13½   8 SIDES 16½-9½   7 SIDES 17½-8¾   8 SIDES 18-7½

FACE OF TONGUE   BACK OF TONGUE

FACE OF BODY   BACK OF BODY

BRACE TABLE
54
54
76. 37

ANGLE CUTS
18 IN.   7 1/2 IN.
8 SIDES

RAFTER TABLE
JACK   9 IN.   9 1/2 IN.
SIDE CUT
1/6 PITCH

ESSEX BOARD MEASURE
6 8/12 FT. B.M.
1 IN. THICK   10 FT

Fig. 11. The front and back views of a typical framing square.

Rafter or Framing Table—This is always found on the body of the square. It is used for determining the length of common valley, hip, and jack rafters and the angles at which they must be cut to fit at the ridge and plate. This table appears as a column six lines deep under each inch graduation from 2 to 18 inches, as seen in Fig. 12A, which shows only the 12-inch section of this table; at the left of the table will be found letters indicating the application of the figures given. Multiplication and angle symbols are applied to this table to prevent errors in laying out angles for cuts.

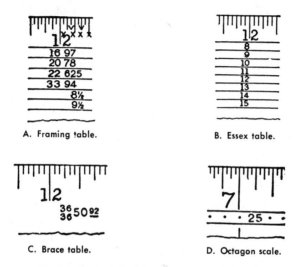

Fig. 12. Typical framing-square markings.

Essex Table—This is always found on the body of the square, as shown in Fig. 12B. This table gives the board measure in feet and twelfths of a foot of boards 1 inch thick of usual lengths and widths. On certain squares, it consists of a table eight lines deep under each graduation, as seen in the figures which represent the 12-inch section of this table.

Brace Table—This table is found on the tongue of the square, a section of which is shown in Fig. 12C. The table gives the length of the brace to be used where the rise and run are from 24 to 60 inches and are equal.

Octagon Scale—This scale is located on the tongue of the square, as shown Fig. 12D, and is used for laying out a figure with eight sides on a square piece of timber. On this scale, the graduations are represented by 65 dots located $5/24$ of an inch apart.

Hundredths Scale—This scale is found on the tongue of the square; by means of a divider, decimals of an inch may be obtained. It is used particularly in reference to brace measure.

Inch Scales—On both the body and the tongue, there are (along the edges) scales of inches graduated in $1/32$, $1/16$, $1/12$, $1/10$, $1/8$, and $1/4$ of an inch. Various combinations of graduations can be obtained according to the type of square. These scales are used in measuring and laying out work to precise dimensions.

Diagonal Scale—Many framing squares are provided with what is known as a diagonal scale, as shown in Fig. 13; one division (ABCD) of this scale is shown enlarged for clearness in Fig. 14. The object of the diagonal scale is to give minute measurements without having the graduations close together where they would be hard to read. In construction of the scale (Fig. 14), the short distance AB is $1/10$ of an inch. Evidently, to divide AB into ten equal parts would bring the divisions so close together that the scale would be difficult to read. Therefore, if AB is divided into ten parts, and the diagonal BD is drawn, the intercepts 1a, 2b, 3c, etc., drawn through 1, 2, 3, etc., parallel to AB, will divide AB into $1/10$, $2/10$, $3/10$, etc., of an inch. Thus, if a distance of $3/10$ AB is required, it may be picked off by placing

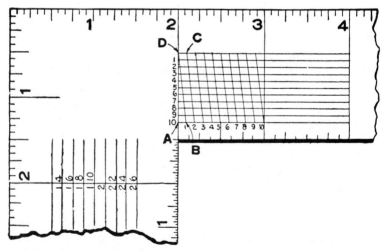

Fig. 13. The diagonal scale on a framing square is used to mark off hundredths of an inch with dividers.

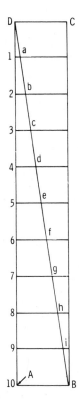

Fig. 14. Section ABCD of Fig. 13, enlarged to illustrate the principle of the diagonal scale.

one leg of the dividers at 3 and the other leg at c, thereby producing $3c = ^3/_{10}$ AB.

Because of the importance of the framing square and the many problems to be solved with it, the applications of the square are given at length in a later chapter.

Combination Square

This tool (Fig. 15), as its name indicates, can be used for the same purposes as an ordinary try square, but it differs from the try square in that the head can be made to slide along the blade and clamp at any desired place; combined with the square, it is also a level and a miter. The sliding of the head is accomplished by means of a central groove in which a guide travels in the head of the square. This permits the scale to be pulled out and used simply as a rule. It is frequently desired to vary the length of the try-square blade; this is readily accomplished with the combination square. It is also convenient to square a piece of wood with a surface and at the same time tell whether one or the other is level, or plumb. The spirit level in the head of the square permits this to be done without the use of a separate level. The head of the square may also be used as a simple level.

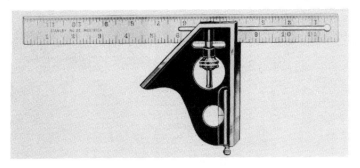

Fig. 15. A typical combination square with a grooved blade, level, and centering attachments.

Because the scale may be moved in the head, the combination square makes a good marking gauge by setting the scale at the proper position and clamping it there. The entire combination square may then be slid along as with an ordinary gauge. As a further convenience, a scriber is held frictionally in the head by a small brass bushing. The scriber head projects from the bottom of the square stock in a convenient place to be withdrawn quickly.

In laying out, the combination square may be used to scribe lines at miter angles as well as at right angles, since one edge of the square head is at 45°. Where micrometer accuracy is not essential, the blade of the combination square may be set at any desired position, and the square may then be used as a depth gauge to measure in mortises, or the end of the scale may be set flush with the edge of the square and used as a height gauge.

The head may be unclamped and entirely removed from the scale, and a center head can then be substituted so that the same tool can quickly be used to find the centers of shafting and other cylindrical pieces. In the best construction, the blade is hardened to prevent the corner from wearing round and destroying the graduations, thus keeping the scale accurate at all times. This combination square combining as it does a rule, square, miter, depth gauge, height gauge, level, and center head permits more rapid work on the part of the carpenter, saves littering the bench with a number of tools each of which is necessary but which may be used only rarely, and tends toward the goal for which all carpenters are striving—greater efficiency. Some of the uses for the combination square are illustrated in Fig. 16.

SLIDING "T" BEVEL

A bevel is virtually a try square with a sliding adjustable blade that can be set at any angle to the stock. In construction, the stock may be of wood or steel; when the stock is made of wood, it normally has brass mountings at each end, and it is sometimes concave along its length. The blade is of steel with parallel sides, and its end is at an angle of 45° with the sides, as shown in Fig. 17. The blade is slotted, thereby allowing linear adjustment and the insertion of a pivot, or screw pin, which is located at the end of the stock. After the blade has been adjusted to any particular angle, it is secured in position by tightening the screw lever on the pivot; this action compresses the sides of the slotted stock together, thus firmly gripping the blade. Fig. 18 illustrates how to set the blade angle.

When selecting a bevel, care should be taken to see that the edges are parallel and that the pivot screw, when tightened, holds the blade firmly without bending it. In the line of special bevels, there are various modifications of the standard or ordinary form of bevel just described. Two of these are shown in Figs. 19 and 20.

CENTER SQUARE

Another useful square which is of fairly recent vintage is the center square (Fig. 20). This works like a protractor and can be used to find the center of any size circle quickly and easily, as well as to determine right angles quickly.

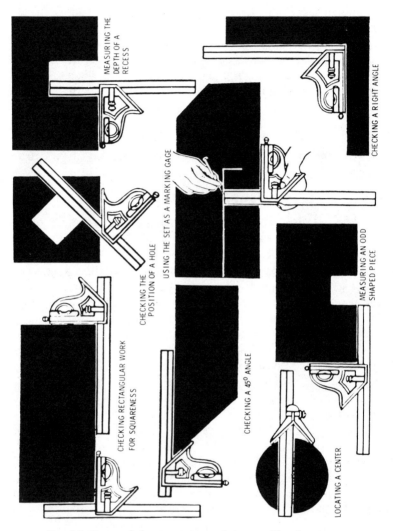

Fig. 16. Some of the many uses of the combination square.

MITER BOXES

The basic device used to make miter cuts is, of course, the miter box (Fig. 21). In its fundamental form it is merely a box with slots cut into it enabling left and right 45° angles to be cut.

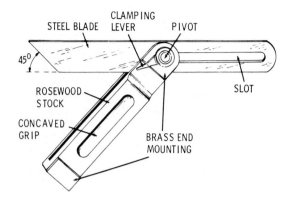

Fig. 17. A sliding "T" bevel with a steel blade, rosewood stock, and brass end mountings. Since the size of a bevel may be expressed by the length of either its stock or its blade, care should be taken to specify which dimension is given when ordering to avoid mistakes.

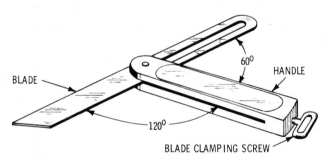

Fig. 18. A sliding "T" bevel. A tool of this type is used to mark and test cutting angles.

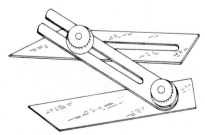

Fig. 19. A double-slot steel bevel. As shown, both the stock and the blade are slotted, thus permitting adjustments that cannot be obtained with a common bevel.

93

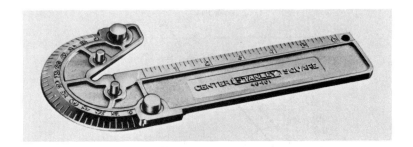

Fig. 20. The center square functions like a protractor and can be used to quickly find the center of a circle or right angles. Courtesy of Stanley.

However, it has come a long way. Today, a wide variety of miter boxes that may be more truly called machines are available. They allow a wide variety of angled cuts, depth of cut adjustment, clamps to hold work, and other desirable features. No matter the miter box, a backsaw is used to make the cuts.

Fig. 21. Today's miter boxes could more properly be called machines. But the basic box still works well. Courtesy of Stanley.

LEVEL

The level is used for both guiding and testing—to guide in bringing the work to a horizontal or vertical position, and to test the accuracy of completed construction. It consists of a long rectangular body of wood or metal that is cut away on its side and near the end to receive glass tubes, which are almost entirely filled with a nonfreezing liquid with a small bubble free to move as the level is moved. A typical level is shown in Fig. 22.

The side and end tubes are at right angles so that when the bubble of the side tube is in the center of the tube, the level is horizontal; when the bubble of the end tube is in the center, the level is vertical. Hence, when the level is placed on a surface, its levelness, or plumbness, can be tested. Levels also come with magnetic edges that can make placement easier.

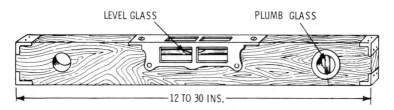

Fig. 22. A typical wooden spirit level with a horizontal and a vertical tube.

PLUMB BOB

The word "plumb" means perpendicular to the plane of the horizon, and since the plane of the horizon is perpendicular to the direction of gravity at any given point, gravity is used to determine vertical lines by the device known as a plumb bob.

This tool consists of a pointed weight attached to a string. When the weight is suspended by the string and allowed to come to rest, as in Fig. 23, the string will be plumb (vertical). The ordinary top-shaped solid plumb bob is problematic because of a too-blunt point and not enough weight. For outside work, the matter of weight is important, since when the plumb bob is used with a strong wind blowing, the excess surface presented to the wind will magnify the error, as shown in Fig. 23. To reduce the

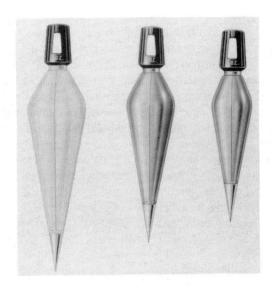

Fig. 23. The solid plumb bob.

surface for a given weight, the bob is bored and filled with mercury. This type of plumb bob is shown in Fig. 24. An adjustable bob with a self-contained reel on which the string is wound is shown in Fig. 25.

Fig. 24. Levels come in a variety of sizes. This small torpedo type is handy for judging the level when the area is not large enough to accommodate a regular-length tool. Courtesy of Stanley.

96

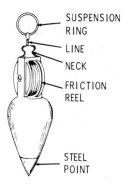

SUSPENSION
RING

LINE

NECK

FRICTION
REEL

STEEL
POINT

*Fig. 25. An adjustable
plumb bob.*

SUMMARY

Guiding and testing tools are invaluable to the carpenter and handyman because without them it would be impossible to mark material with the precision that is often demanded. A good variety of these tools should be in every toolbox, and should be used frequently.

One of the more important guiding and testing tools is the straightedge, which can be of metal or wood and can be used to guide the pencil or scriber. It also comes in a variety of lengths.

The try square is used to check for right-angle cuts on any straightedged material. There are various types of try squares, such as double try squares, combination try squares, and try-and-miter squares.

A framing square (sometimes called steel square) is a square with framing tables and various other scales. It is adapted especially for use in house framing. The features that make this square such a valuable tool are the table of rafter measurements for common, valley, hip, and jack rafters, and the angles at which they must be cut to fit the ridge and plate.

The miter box is a tool used to guide the saw in cutting material, generally at 45° and 90° angles. A miter box has at least two 45° angles for cutting right and left miters but many miter boxes allow other angle cuts.

The level is a tool used to bring the work to either a true horizontal or vertical position. By holding the level on a surface that is horizontal or vertical, the tool itself may be checked for accuracy.

REVIEW QUESTIONS

1. What is a straightedge?
2. Explain the use of the plumb bob and the level.
3. Name a few of the common types of squares.
4. What is a shooting board?
5. Why does a miter box have two 45° angles?

CHAPTER 8

Marking Tools

In good carpentry and joinery, a great deal depends on the accuracy achieved in laying out the work. The term "laying out" means the operation of marking the work with a tool, such as a pencil or scriber, so that the various centers and working lines will be set off in their proper relation. These lines are followed by the carpenter in cutting and other tooling operations necessary to bring the work to its final form.

In laying out, the guiding tools just described in Chapter 7 are used to guide the pencil or scriber; the measurements are made by the aid of the measuring devices described in Chapter 9.

According to the degree of precision required in laying out, the proper marker to use is:

1. For extremely rough work—the chalk box and reel or the carpenter's pencil with rectangular lead.
2. For rough work—the lead pencil with round lead.
3. For semirough work—the scratch awl.
4. For precision work—the scriber or knife.

For efficiency, a good degree of common sense should be used in deciding which marker to use. Thus, it would be self-defeating to use a machine-hardened steel scriber with a needle point to mark off rafters, or to use a carpenter's pencil with "an acre of soft lead" on the point to lay out a fine dovetail joint.

CHALK BOX AND LINE

The chalk line, which is shown in Fig. 1, is to mark a long straight line between two points that are too far apart to permit the use of a square or straightedge.

The chalk box is usually constructed of aluminum or plastic. Inside the box is a reel that is fitted with a lightweight string or cord; the box has powdered chalk in it. To use it, the line is stretched between two points; when the string is taut, it is pulled up and released, thus leaving a chalk line on the surface. Note the right way and the wrong way to use the chalk line, as shown in Fig. 2.

CARPENTER'S PENCIL

The conventional carpenter's pencil, which is rectangular in cross section, is considerably larger than an ordinary pencil. The idea in making the lead this shape is to permit its use on rough lumber without too frequent sharpening and to give a well-

Fig. 1. A chalk box with 100 feet of string.

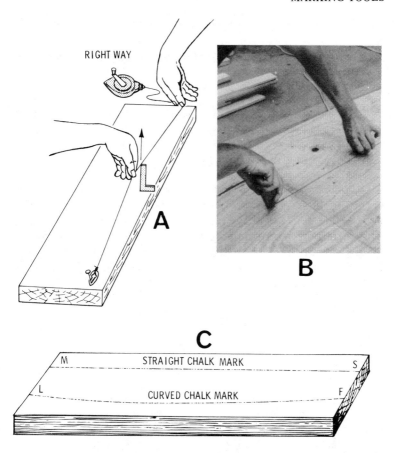

RIGHT WAY

A

B

C

M   STRAIGHT CHALK MARK   S

L   CURVED CHALK MARK   F

Fig. 2. The right way to use the chalk box. When pulling up the line, always do so in a direction that is at right angles with the board. If the chalk line is pulled straight up, as in A, a straight chalk mark MS will be obtained; if the line is pulled up to one side, a curved line LF will be produced.

defined, plainly visible line. Because of the thickness of the line, the carpenter's pencil is not intended for fine work but is used principally for marking boards, etc., that are to be sawed.

Fig. 3 shows a section of the carpenter's pencil. When marking with the pencil, the mark must be made in the direction of the long axis of the lead, as shown in Fig. 4B, and not as in Fig.

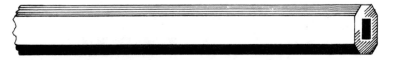

Fig. 3. Section of a typical carpenter's pencil.

4C. The proper method of sharpening the pencil, as shown in Fig. 4A, should be noted.

ORDINARY PENCIL

Ordinary pencils with thin leads are of course also used in carpentry. Since the lead *is* smaller than that of the carpenter's pencil, it produces a finer marking line and is used on smooth surfaces where more accurate marking is required than can be obtained with the carpenter's pencil. When using, the best results are obtained by twisting the pencil while drawing the lines so as to retain the conical shape the lead gets when sharpened.

MARKING OR "SCRATCH" AWL

This tool consists of a short piece of round steel that is pointed at one end with the other end permanently fixed in a

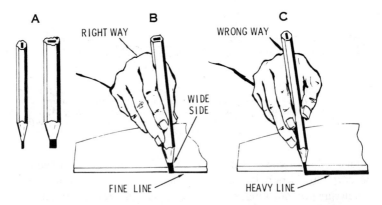

Fig. 4. The right and wrong ways to use the carpenter's pencil; A, side and end views of a carpenter's pencil; B, a fine line is obtained with the long side of the lead turned in the direction of the straightedge; C, a wide, undefined line is produced when the pencil is used in this position.

convenient handle, as shown in Fig. 5. A scratch awl is used in laying out fine work where a lead pencil mark would be too coarse for the required degree of precision.

SCRIBER

A scriber is a tool of extreme precision, and while intended especially for machinists, it should be in the tool kit of carpenters, mechanics, and do-it-yourselfers who are engaged in very fine work.

A scriber is a hardened steel tool with a sharp point designed to mark extremely fine lines. The most convenient form of scriber is the pocket, or telescoping, type, shown in Fig. 6; the construction renders it safe to carry in the pocket.

COMPASS AND DIVIDERS

The compass is an instrument used for describing circles or arcs by scribing. It consists of two pointed legs that are hinged

Fig. 5. An ordinary scratch awl with a forged blade and a hardwood handle.

Fig. 6. A telescoping scriber in the open and closed positions.

103

firmly by a rivet so as to remain set in any position by the friction of the hinged joint. The usual form of carpenter's compass is shown in Fig. 7. It should not be used in the place of dividers for dividing an arc or line into a number of equal divisions because it is not a tool of precision.

The difference between dividers and compasses is that the dividers are provided with a quadrantal wing projecting from one of the two hinged legs through a slot in the other leg. A setscrew on the slotted leg enables the instrument to be securely locked to the approximate dimension and adjusted with precision to the exact dimension by a screw at the other end of the wing. A spring pressing against the wing holds the leg firmly against the screw. Its general appearance is shown in Fig. 8. Because of the wing, the tool is frequently called winged dividers.

Fig. 7. A typical compass.

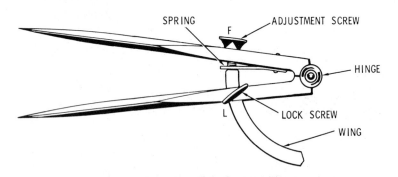

Fig. 8. Winged dividers for describing and dividing arcs and circles. When the dividers are locked in the approximate setting by lock screw L, the tool can be set with precision to the exact dimension by turning adjustment screw F, against which the leg is firmly held by the spring to prevent any lost motion.

SUMMARY

Accuracy in carpentry work depends on the correct use of good tools. In layout work, the guiding tools are used to guide a pencil or scriber.

The conventional carpenter's pencil has a rectangular lead that is considerably larger than the lead in an ordinary pencil. With this design, the pencil can be used frequently without sharpening the lead.

The scratch awl is a short piece of round steel that is pointed at one end with the other end permanently fixed in a convenient handle. A scratch awl is used in laying out fine work where a pencil would be too coarse for the required precision.

The compass or divider is an instrument used for describing circles or arcs. It is designed with two pointed legs that are hinged at one end.

REVIEW QUESTIONS

1. What is the difference between a divider and a compass?
2. Why is a carpenter pencil lead rectangular?
3. What is the purpose of the scratch awl?
4. What marking tool would be used for precision work?
5. What is a chalk line and how is it used?

CHAPTER 9

Measuring Tools

A number of measuring tools are considered essential, and some are considered very helpful, for the carpenter. The handyman, too, should give these items careful buying consideration.

RULES

The most basic tool for the carpenter is the ruler, or rule. Most carpenters favor the 12-foot-long steel tape. It enables them to take both short and long measurements with accuracy.

Folding Wood Rule

Another rule—an old favorite and deservedly so—is the folding wooden rule. This commonly comes 6 feet long and in one-foot increments that are hinged together and fold out for use. Such a ruler is invaluable because it permits one-hand measuring of something: It has the rigidity required for laying it across large areas.

A variation on the basic wood rule is the wood rule with an extension that pulls out of one end. This is good when you are measuring an enclosed area and the regular segments of the folding rule are too large for it. Just slide the extension out to measure the last few inches.

Other Rules

The bench rule is another rule that is handy. It is mounted on the edge of a bench and things are held against it for measuring.

A 6-inch caliper rule is also good. It has a hooklike part that

Fig. 1. Most carpenters favor 12-foot steel tape. It can be used for long or short measuring jobs. Courtesy of The American Plywood Assn.

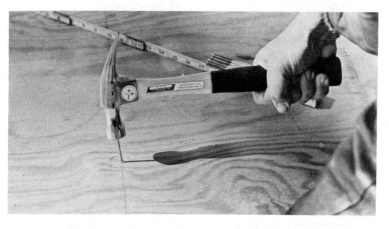

Fig. 2. The folding rule is another old favorite. One advantage of the rule is that it can be used with one hand. Courtesy of Vaughn & Bushnell.

Fig. 3. The folding rule is also good for stretching across an area. It has the required rigidity.

slides in a groove along the rule. The outside caliper rule is excellent for checking the thickness of round stock. You can also get caliper rules for checking the widths of slots and grooves. Such rules are known as *inside* caliper rules.

The big idea when buying a rule, as with other kinds of tools, is to buy quality. A saw that cuts in an errant way may be tolerated for the duration of a job, but a measuring tool that isn't accurate can turn a job into a disaster.

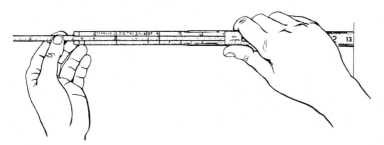

Fig. 4. The folding rule with extension is good when regular segments are too big for measurement.

USING RULES

The axiom when using any kind of measuring device is to "measure twice, cut once." Another notion, not so well known, is to use the same brands of rules if you are working with someone. What constitutes 12 feet to one manufacturer may not mean the same thing to another, even though both produce quality

109

tools. If two brands of rules, then, differ by only $^1/_{32}$ inch per foot, this may not mean much in a short measuring job, but it can spell big problems as the footage goes up.

There are a number of other tips to follow:

When measuring anything, hold the rule on its edge so that the measuring marks are in contact with the work. Make the mark with a sharp knife or pencil.

If you are measuring the end of a board, use a forefinger to guide one end of the rule into alignment with one side of the board, and a thumbnail to fix the exact measuring mark on the other side.

If you want to make a line parallel to the edge of a board, hold the rule as shown in Fig. 7, guiding the rule along with the pencil in the position shown.

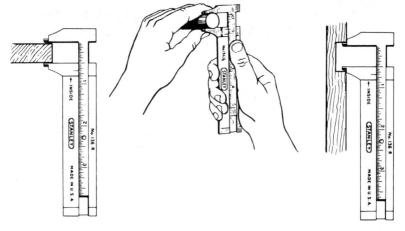

Fig. 5. An outside caliper (left and center) can be used to measure the outside of square or round objects. An inside caliper (right) is for measuring widths of grooves and the like.

LUMBER SCALES

When estimating lumber, a lot of time is saved in the lumberyard by using a lumber scale from which the board feet measure may be read off directly. This scale gives an approximate result. When using the scale, it is customary to read to the nearest figure and when there is no difference to alternate be-

Fig. 6. To make a mark with a rule on the end of board, place rule up so that marks are on board, then mark.

Fig. 7. To measure end of board, use your forefinger to guide one end of the rule flush, then use thumb as shown to fix mark.

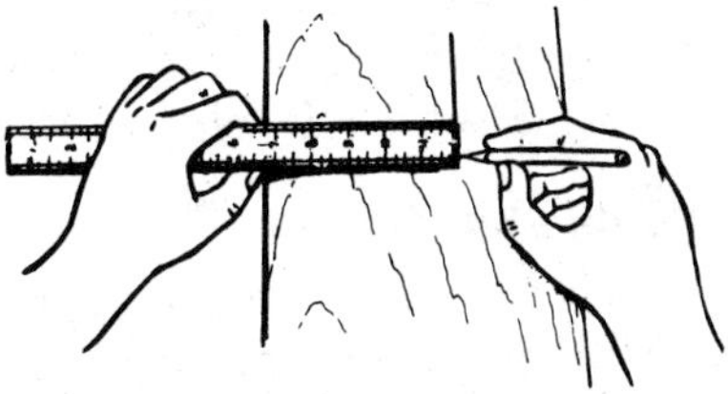

Fig. 8. To make a mark parallel to edge, hold the rule as shown and place pencil on end of rule; pull along.

tween the lower and higher figures on different boards. Fig. 9 shows a board scale graduated for boards of 12-, 14-, and 16-foot lengths and the method of using the scale.

There are many types of lumber scales in use in various sections of the United States and Canada.

Spring Steel Board Rule

This rule is made of tempered spring steel so that it will bend to the board and, when released, will return exactly straight. It is provided with a wood handle and a leather slide for handling the rule at any part of the blade. Types—3-tier, 3½-foot inspector's rule; 3-tier, 3-foot board rule; 3-tier, 2½-foot sorting rule. Markings—all three types are marked on one side to measure 8, 10, and 18 feet; the opposite side is marked 12, 14, 16, or 18, 20, and 22 feet.

MARKING GAUGES

Tools of this type are used to mark a piece of wood that is to be sawed or otherwise tooled. There are several types of marking gauges, such as:

1. Single bar.
2. Double bar.
3. Single bar with slide.
4. Butt.

Single-Bar Gauge

This is used for making a single mark, such as for sawing. It consists of a bar with a scriber, or pin, at one end, provided with a scale that is graduated in inches and sixteenths. The bar passes through a movable head that may be clamped at any distance from the scriber point, as shown in Fig. 10.

Double-Bar Gauge

This type of gauge is designed especially for mortise marking. There are two independent bars working in the same head. One

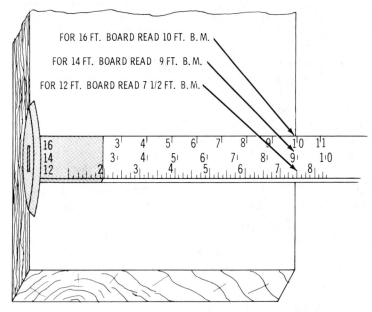

Fig. 9. A typical board rule that is used to measure lumber in board-measure units. To use the rule, place the head of the rule against one edge of the board, and read the figure nearest the other edge (width) of the board in the same line of figures on which the length is found. This reading will give the number of feet board measure for the piece of lumber being measured.

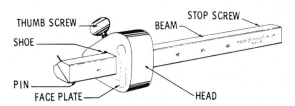

Fig. 10. A single-bar marking gauge. It is fitted with a head, faceplate, and thumbscrew and is provided with a scale that is graduated in inches and 16ths.

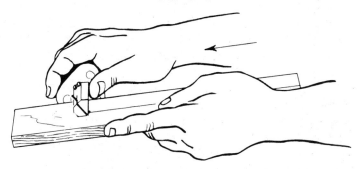

Fig. 11. The method of using a marking gauge. When setting the gauge, use a rule, unless it is certain that the scriber point is located accurately with the graduations on the bar. When marking, the gauge should be held as indicated; the face of the head is pressed against the edge of the board. Care must be taken to keep it true with the edge so that the bar will be at right angles with the edge, and the line scriber will be at the correct distance from the edge. The line is usually scribed by pushing the gauge away from the worker. Always work from the face side, as shown.

pin is affixed to each bar. After setting the bars for the proper marking of the mortise, one side is marked with one bar, and the gauge is then turned over for marking the other side. The construction is shown in Fig. 12.

Slide Gauge

One objection to the double-bar gauge is that two operations are required that can both be performed with a slide gauge in one operation. As shown in Fig. 13, the underside of the bar is provided with a flush slide having a scriber B at the end of the slide, with another scriber A at the end of the bar. These two scribers, when set to the required distances from the head, mark both sides of the tenon or mortise to the size required with just

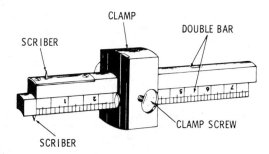

Fig. 12. A double-bar marking gauge. It is used for marking a given distance between two parallel lines and a given distance from the edge of a board. As shown in the illustration, each bar has a setscrew for clamping it in any desired position.

one stroke. On the upper side, there is one scriber C for single marking.

Butt Gauge

When hanging doors, there are three measurements to be marked:

1. The location of the butt on the casing.
2. The location of the butt on the door.
3. The thickness of the butt on the casing.

A butt gauge is a type of gauge having three cutters, which are purposely arranged so that no change of setting is necessary

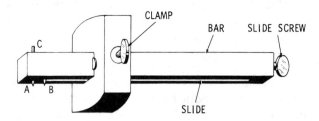

Fig. 13. A slide marking gauge. The bar has a scriber (C) on the upper side for single marking and a scriber (A) on the lower side which, with the scriber (B) on the slide, works flush in the bar. The distance between scriber points A and B is regulated by the slide screw at the end of the bar.

Fig. 14. The method of using a slide gauge when marking a mortise.
Note that the marks (M and S) for the sides of the mortise can both be
scribed in one operation.

when hanging several doors. In reality, these tools comprise
rabbet gauges, marking gauges, and mortise gauges of a scope
sufficient for all door trim, including lock plates, strike plates,
etc.

Fig. 15 shows a typical butt gauge. The cutters are mounted
on the same bar and are set by one adjustment with the proper
allowance for clearance. When casings have a nailed-on strike
instead of being rabbeted, a marking gauge that will work on a
ledge as narrow as ⅛ inch is required; in this case, the same
distance is marked from the edge of the casing and from the edge
of the door that is not engaged when closing. Certain gauges can
be used on such work; one cutter marks the butt, and one cutter
marks its thickness. Other gauges are made so that they can be
used as inside or outside squares for squaring the edge of the
butt on either the door or the jamb.

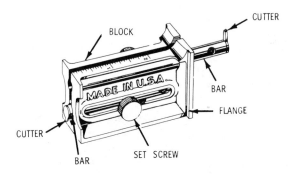

Fig. 15. A typical butt gauge. It is used to mark the location of the butt on casings and doors. Three separate cutters, one for each dimension, eliminate changing the setting when more than one door is hung. It is also used as a marking and mortising gauge and an inside and outside square for squaring the edge of a butt on the door and jamb; it is graduated in 16ths for 2 inches.

SUMMARY

The most common measuring tool known in any type of work is the rule. There are many different types of so-called tools that fall in this category, but in this type of work it is referred to as the carpenter's rule. One familiar type is the 6-foot folded wooden rule. The most popular rule is the 12-foot steel tape.

Marking gauges are used to mark pieces of wood that are to be sawed. There are several types of marking gauges.

REVIEW QUESTIONS

1. What type of rule is most popular?
2. What is a marking gauge? Explain how it works.
3. How do you use a rule to mark a line parallel to a board?

CHAPTER 10

Holding Tools

Sometimes holding tools such as clamps, vises, and horses get lost in the shop shuffle. They shouldn't. A good selection of holding tools can make home and shop tasks considerably easier, and there are a variety of jobs that cannot be done properly without them. Following is a consideration of the essentials.

The workbench, considered broadly with its attachments, may be called the main holding tool, and unless this important part of the equipment is constructed amply substantial and rigid, it will be difficult to do good work. The workbench and its attachments have already been described. When installing a bench, it should be properly anchored or fastened to the wall of the room very solidly—indeed, as if it were part of the wall.

Holding tools may be generally classed as supporting tools and retaining tools. When marking or sawing, it is usually only necessary to support the work by placing it on the bench or on sawhorses; however, in planing, chiseling, and some nailing operations, the work must not only be supported but held rigidly in position.

"HORSES" OR TRESTLES

"Horses" or trestles are used in various ways to simply support the work when it is of such large dimensions that the bench cannot conveniently be used, especially for marking and sawing planks. No shop equipment list is complete without a pair of sawhorses. A sawhorse, as usually made, consists of a 3- or 4-foot length of 2″ × 4″ or 2″ × 6″ stock for the cross beam, with a pair of 1″ × 3″ or 1¼″ × 4″ legs at each end, depending on the expected weight of the work.

The height of the sawhorse is usually 2 feet. The general construction of sawhorses is shown in Fig. 1. Note that the legs are inclined outward both lengthwise and crosswise, and a problem arises as to how to determine the length of the legs having this double inclination for a given height of sawhorse; the solution is shown in Fig. 2, and the method of obtaining the angular setting for the bevel to scribe the mortise is illustrated in Fig. 3.

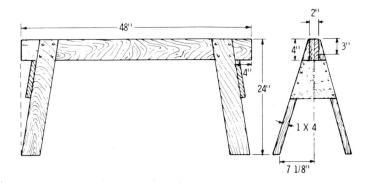

Fig. 1. The side and end views of a typical carpenter's sawhorse whose dimensions are suitable for general use.

CLAMPS

The quality of a particular job is largely the result of how securely and well parts are held together while the glue is drying. In the old days, many carpentry joints depended on stresses from the way the joints were cut and joined, but today the job is done by clamps.

There are quite a few different kinds of clamps available, some generally used, some for more specific uses. Following is a roundup:

C Clamp—These come with jaws that open to 12 inches wide. As the name suggests, they are in the shape of the letter C, and are very useful for clamping a wide variety of items. Some C clamps come with shielded screws to protect against hammer blows.

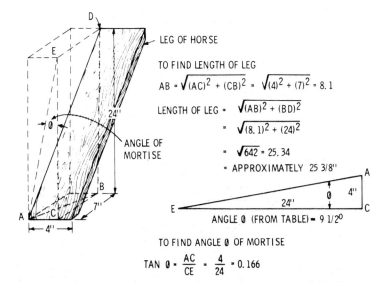

LEG OF HORSE

TO FIND LENGTH OF LEG

$AB = \sqrt{(AC)^2 + (CB)^2} = \sqrt{(4)^2 + (7)^2} = 8.1$

$$\text{LENGTH OF LEG} = \sqrt{(AB)^2 + (BD)^2}$$
$$= \sqrt{(8.1)^2 + (24)^2}$$
$$= \sqrt{642} = 25.34$$
$$= \text{APPROXIMATELY } 25\ 3/8''$$

ANGLE Ø (FROM TABLE) = 9 1/2°

TO FIND ANGLE Ø OF MORTISE

$$\text{TAN } \emptyset = \frac{AC}{CE} = \frac{4}{24} = 0.166$$

Fig. 2. The method of finding the length of a sawhorse leg and the angle, or inclination of side, of the mortise for the leg. To find ϕ by calculation, a table of natural trigonometric functions is necessary.

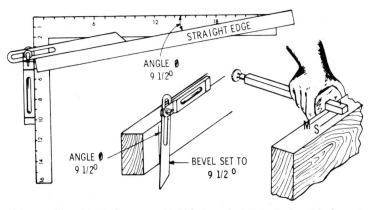

Fig. 3. The method of setting the bevel angle (ϕ) with the aid of a square and a straightedge. Place the straightedge on the square so that one side of the right triangle thus formed will be 24 inches (height of the sawhorse) and the other side will be 4 inches (distance to edge of leg from end of beam). Place the blade of the bevel against the straightedge, and place the stock against the side of the square; clamp the bevel to this angle, which is the proper slope for the side of the mortise.

Fig. 4. C clamps in action. They are very useful. Courtesy of The American Plywood Assn.

Deep Throat Type—Closely aligned to the regular C clamp is the deep throat type. It is good when you need to slip the clamp over the edge of something to clamp it someplace on the interior where other clamps would have difficulty reaching.

Edge Clamp—There are a few kinds of edge clamps, one of which is basically a C clamp with another screw set in at a right angle to the jaw opening.

Pipe Clamp—Separate fittings that comprise the clamp are bought separately and are fastened to ½- or ¾-inch iron pipe of whatever length is desired. This type of clamp can be used to clamp very deep work.

Steel Bar Clamp—This resembles the bar clamp (see Fig. 5). It is available with openings from 2 feet to 8 feet, 5 inches deep.

Hand Screw—This clamp is best for holding items where the

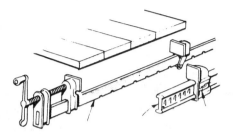

Fig. 5. Bar clamp, good for wide work. Courtesy of Hand Tool Institute.

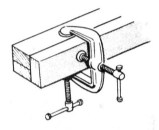

Fig. 6. Edge, or corner, clamp, good for securing stock to edges. Courtesy of Hand Tool Institute.

faces of the pieces are not parallel. Other clamps would slip off, but the hand screw can be adjusted to hold uneven work and round things 2 to 17 inches deep.

Spring Clamp—These are like having an extra hand. They are like oversized clothespins and can be used to apply a good deal of pressure.

Band Clamp—This is for clamping large, irregularly spaced items. There is a strip of nylon or canvas webbing 1 to 2 inches wide and 15 feet long. The ratchet head is tightened securely by wrench or handle.

VISES

Vises are usually permanently attached to a workbench or table. There are a number of useful styles of vises.

Woodworker's Vise

The woodworker's vise has two parallel flat jaws varying in size from 3 × 6 inches to 4 × 10 inches. Jaws open from 6 inches to up to 12 inches on professional models. The best position for

121

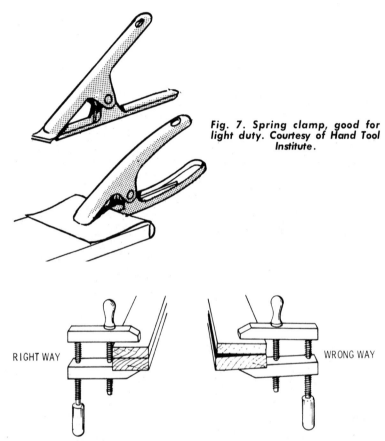

Fig. 7. Spring clamp, good for light duty. Courtesy of Hand Tool Institute.

RIGHT WAY   WRONG WAY

Fig. 8. Right and wrong ways to use a hand screw. First, set the jaws to approximately the size of the material to be clamped. When placing the hand screw on the work, keep the points of the jaws slightly more open than the outer ends. Final adjustment of the inside screw will then bring the jaws exactly parallel, which is the proper position for clamping parallel work. Of course, if the work itself is wedge-shaped, the clamp jaws should conform so that equal pressure is applied at all points of contact. Since the screws are made of wood instead of iron or steel, proper adjustment must be used with respect to the applied pressure.

a woodworker's vise is to have the tops of the jaws flush with the workbench top.

One feature that signifies a better vise is a fast-acting screw that allows the front jaw of the vise to be quickly positioned in a

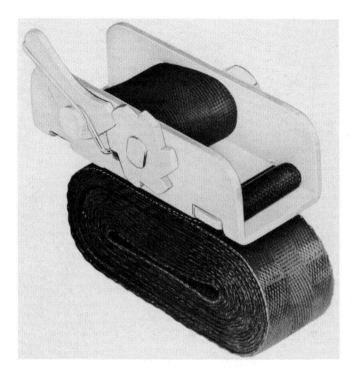

Fig. 9. A band clamp which is used to clamp irregularly shaped objects.

gripping position prior to a tightening action, something accomplished by rotating a sliding pin handle, which turns a screw that applies the force. The jaws are never barefaced; wood or composition inserts are installed to protect the workpiece.

The woodworker's vise can be used to hold a wide variety of things, from boards to sheets of plywood.

One other good feature of the woodworker's vise is a spring-loaded "dog" in the front jaw. This can be raised to butt against one side of a very wide workpiece so that piece is much wider than the normal jaw opening can be held.

Bench Vise

The bench vise has either a fixed base that is bolted to the bench top or a swivel base that allows the vise to be turned 220°

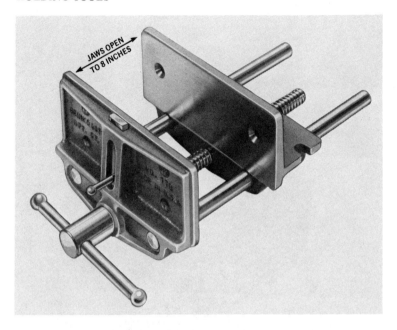

Fig. 10. Woodworker's vise. Courtesy of Brink & Cotton.

to 360°. Whichever, the main screw can be turned to open the vise as required. Jaws may be opened from 3 to 6 inches depending on the vise. Construction varies; vises are made of gray iron, malleable iron, or steel.

Machinist vises, which get more rugged use, are usually made of steel or malleable iron and are more expensive than bench vises.

Good bench vises usually include a set of pipe jaws for working pipe; better vises also have hardened steel and serrated jaw inserts which can be used for holding pipe.

Vises should be mounted so that the jaws extend over the edge of the bench. This way, long objects can be held vertically without interference.

Clamp-On and Sawhorse Vises

Two other useful vises are the clamp-on and the sawhorse. Clamp-on vises are basically bench vises, but with jaws that

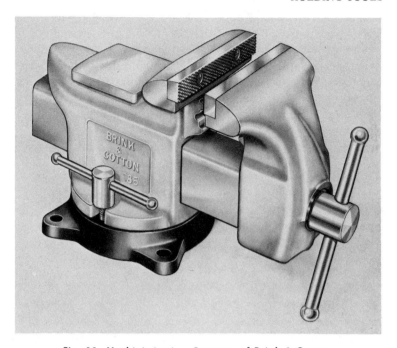

Fig. 11. Machinist's vise. Courtesy of Brink & Cotton.

open only to 3 inches. As the name suggests, they clamp onto the bench. They are commonly used for light work.

The sawhorse vise is a variation of the clamp-on bench vise, but is designed to be clamped to a sawhorse. The sawhorse vise is L-shaped and is good for holding a board or piece of plywood in a vertical position as work is done on it.

SUMMARY

Holding tools are essential parts of workshop equipment. Many tooling operations require that the work be held rigid. As described in Chapter 5, the workbench can be considered a holding tool and should be anchored to the floor or wall in order to prevent movement.

Horses or trestles are a simple way to support work when large dimensions are being used. This type of labor-saving device is generally made from $2'' \times 4''$ or $2'' \times 6''$ stock for the cross beam, with a pair of $1'' \times 3''$ or $1'' \times 4''$ legs at each end.

An essential tool for any workbench is a vise. There are various types, such as woodworker's, bench, and sawhorse. The bench vise is not always a convenient tool to carry to a particular job, so in such cases clamps are used. Various types of clamps are available including C clamps, miter clamps, and bar clamps.

REVIEW QUESTIONS

1. What are some advantages in using a vise?
2. What is a sawhorse?
3. Why are clamps so important?
4. What is a chain clamp and when would it be used?
5. What is a hand screw?

CHAPTER 11

Toothed Cutting Tools

In almost all carpentry jobs, after the work has been laid out with guiding, marking, and measuring tools, and supported or held in position by a holding tool, the first cutting operation will be performed by a toothed tool. The most important of these kinds of tools is the saw. Since sawing is hard work, the carpenter should know not only how to saw properly but also how to keep the saw in prime condition.

SAW TYPES

Of course, a number of saws are available, and they can be characterized in a variety of ways: in terms of type of blade, use, back reinforcement, and others. But the typical handsaw is the crosscut type, meaning that its blade is designed to cut wood across the grain. It is commonly 26 inches long.

Fig. 1. A typical handsaw used in various woodcutting operations. The coarseness or fineness of a saw is determined by the number of teeth per inch. A coarser saw, properly set, is preferred for fast work on soft and green wood, whereas a finer saw is suitable for smooth, accurate cutting and for dry, seasoned wood. Ripsaws commonly have 5½ to 6 teeth per inch, and crosscut saws have 7 or 8 teeth per inch.

127

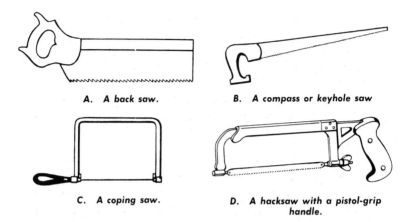

A. A back saw.

B. A compass or keyhole saw

C. A coping saw.

D. A hacksaw with a pistol-grip handle.

Fig. 2. Four popular types of saws.

A handsaw that looks very much like the crosscut, differing in the design of the teeth, is the ripsaw. This saw is geared for cutting with the grain, or ripping a board. Of course, most carpenters and handymen will use a power tool whenever possible for operations that involve extensive cutting.

Handsaws—crosscut or rip—will have a handle of wood or plastic and vary in length from 14 to 30 inches; in the smaller sizes, such a saw is called a panel saw. Table 1 indicates the lengths available in these saws.

Table 1. Saw Sizes

	Panel						Hand	Rip	
Size Inches	14	16	18	20	22	24	26	28	30

Other Saws

A variety of other saws are very useful, designed to do jobs a regular handsaw cannot, or cannot do easily.

Backsaw (Fig. 2A)—a thin crosscut saw with fine teeth that is stiffened by a thick steel web through the entire length of the blade along the back edge. One popular size has a 12-inch blade length with 14 teeth per inch. It is used for making joints and in fine woodworking operations where great accuracy is desired.

Compass, or Keyhole Saw (Fig. 2B)—has a small narrow blade with a pistol-grip handle and is commonly used for cutting along circular curves or lines in fine or small work.

Coping Saw (Fig. 2C)—also used for cutting curves or circles in thin wood; it consists of a small narrow blade that is inserted in a sturdy metal frame in a manner similar to that used in the hacksaw.

Hacksaw (Fig. 2D)—although used primarily for cutting metal, is a popular tool in any woodworking shop. There are two parts to a hacksaw—the frame and the blade. Common hacksaws may have either adjustable or solid frames, although the adjustable frame is generally preferred. Hacksaw blades of various types are inserted in these adjustable frames for different kinds of work; the blades vary in length from 8 to 16 inches. The blades are usually ½ inch wide and have from 14 to 32 teeth per inch. A hole at each end of the blade allows the blade to be hooked to the frame; a wing nut on one end regulates the blade tension, thus permitting various types of cutting actions.

Still other saws with special uses are the dovetail saw, which is geared for cutting dovetail joints, and the drywall saw for cutting material of the same name. Of most recent vintage is the rod saw, which has a narrow blade coated with tungsten carbide particles and can be used to cut very hard materials such as glass and ceramic tile.

Saw Teeth

The cutting edge of a handsaw is a series of little notches, all of the same size. On a crosscut saw, each side of the tooth is filed to a cutting edge like a little knife, as illustrated in Fig. 3. On a ripsaw, each tooth is filed straight across to a sharp square edge like a little chisel, as shown in Fig. 4.

Set

The set of a saw is the distance that the teeth project beyond the surface of the blade. The teeth are "set" to prevent the saw from binding and the teeth from choking up with sawdust. In setting, the teeth are bent alternately, one to one side and the next to the other side, thus forming two parallel rows, or lines, along the edge.

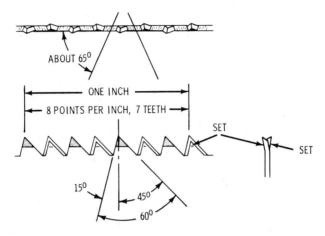

Fig. 3. The side and tooth-edge views of a typical crosscut saw. This saw is used for cutting across the grain and has a different cutting action than that of the ripsaw. The crosscut saw cuts on both the forward and backward strokes.

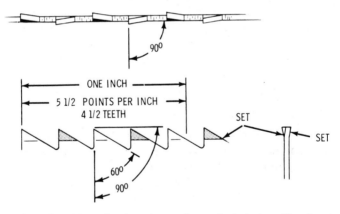

Fig. 4. The side and tooth-edge views of a typical ripsaw. The ripsaw is used for cutting with the grain. Cutting is done only on the forward stroke.

Action of the Crosscut Saw

While each crosscut tooth resembles a little two-edged knife, it cuts quite differently. In early times, it was discovered that a knife blade must be free from nicks and notches to cut well.

130

Then it could be pushed against a piece of wood, and a shaving could be whittled off. At about the same time, it was noticed that if the nicked knife were drawn back and forth across the wood, it would tear the fibers apart, producing sawdust.

Fig. 5 shows a good way of using the crosscut saw. The saw is solidly gripped as is the board, and the cutting line can be clearly seen. An angle of approximately 45° should be maintained between the saw and the face of the work.

The set of crosscut teeth makes them lie in two parallel rows. A needle will slide between them from one end of the saw to the other. When the saw is moved back and forth, the points, especially their forward edges, sever the fibers in two places, leaving a little triangular elevation that is crumbled off by friction as the saw passes through. New fibers are then attacked, and the saw drops deeper into the cut.

Fig. 5. The approved method of crosscutting.

Action of the Ripsaw

The teeth of the ripsaw are a series of little chisels set in two parallel rows that overlap each other. At each stroke, the sharp edge chisels off a little from the end of the wood fibers, as shown in Fig. 6. The teeth are made strong with an acute cutting angle, but the steel is softer than that of a chisel to enable the teeth to be filed and set readily. Fig. 7 illustrates the proper position for cutting with a ripsaw.

Angles of Saw Teeth

The "face" of each crosscut tooth is slightly steeper than the back, thereby making an angle with the line of the teeth of ap-

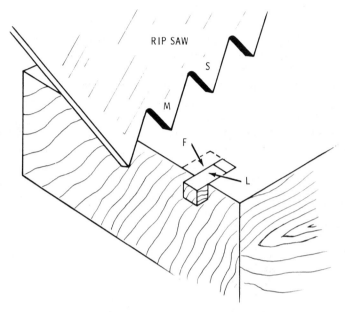

Fig. 6. Action of the ripsaw. When the first tooth is thrust against the wood at an angle of approximately 45°, it chisels off and crowds out small particles of wood. Thus, tooth M will start the cut and take off piece L; tooth S will take off piece F, and so on.

Fig. 7. The proper position when using the ripsaw.

proximately 66°. The compass teeth lean still further at an angle of 75°. The ripsaw face is at right angles (90°) to the line of the teeth. Its cutting edge is at right angles to the side of the blade. The angle of each tooth covers 60°. These angles are shown in Fig. 9.

Fig. 8. To saw, grasp the wood with the left hand, and guide the saw with the thumb. Hold the saw lightly, and do not press it into the wood; simply move it back and forth, using long strokes.

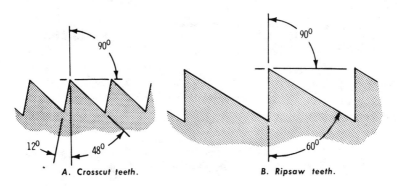

A. Crosscut teeth.   B. Ripsaw teeth.

Fig. 9. Angular proportions for crosscut and ripsaw teeth.

133

FILES AND RASPS

By definition, a file is a steel instrument having its surface covered with sharp-edge furrows or teeth and is used for abrading or smoothing other substances, such as metal and wood. A rasp is a coarse file and differs from the ordinary file in that its teeth consist of projecting points instead of grooves cut across the face of the file.

Files are used for many purposes in woodworking. Fig. 10 shows a variety of them. The taper file is adapted for sharpening hand, pruning, and buck saws. The teeth of the mill file leave a smooth surface. They are particularly adapted to filing and sharpening mill saws and mowing- and reaping-machine cutters. Rasps are generally used for cutting away or smoothing wood or for finishing off the rough edge left in a circular hole that has been cut with the keyhole saw. The ordinary wood rasp is rougher or coarser than that used by cabinetmakers. Wood files are usually tempered to handle lead or soft brass.

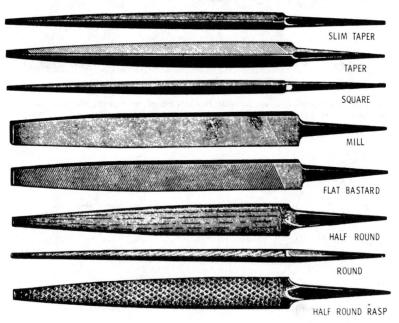

SLIM TAPER

TAPER

SQUARE

MILL

FLAT BASTARD

HALF ROUND

ROUND

HALF ROUND RASP

Fig. 10. Various types of files and rasps.

In using large rasps or files, whether for wood or metal, the work should be held in the vise or otherwise firmly fixed, because it is desirable to use both hands when possible. The handle of the tool should be grasped by one hand while the other hand is pressed, but not too heavily, on the end or near the end of the blade so as to lend weight to the tool and add to its powers of abrasion.

SURFORM TOOLS

A tool that is somewhere between a rasp and a file and is excellent for removing everything from wood to soft metal is the Surform tool from Stanley. It comes in a variety of shapes and has a blade that looks like a cheese grater. As the blade is passed over the material, the chips pass through the holes.

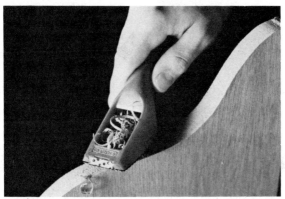

Fig. 11. The Stanley Surform tool comes in a variety of shapes for different cutting jobs. Its blade looks like a cheese grater. Courtesy of Stanley.

SANDPAPER

While not a tool in the strictest sense, sandpaper most certainly is a cutter, and a knowledge of it is well advised for anyone working with wood.

In reality, sand is never used in making sandpaper, and the backing material is often not paper. It is paper, paper/cloth combinations, and various fibers. Properly speaking, the name for sandpaper is coated abrasive.

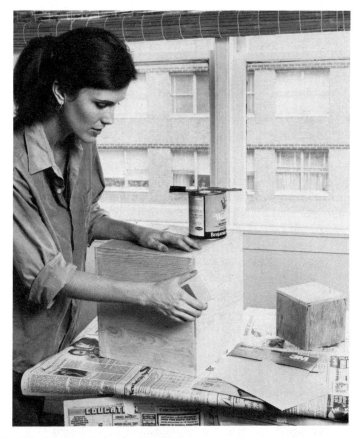

Fig. 12. Sandpaper isn't really sandpaper, nor is it a cutting tool in the ordinary sense, but it does cut and is essential in woodworking and carpentry. A block or sandpaper holder works well. Courtesy of 3M.

Sandpaper is characterized in a number of ways: by the so-called retail system, where there are word descriptions of its uses; by numbers; by the industrial system, which is widely used; and by the old system, which is also numbers. Table 2 breaks it all down.

The grit on sandpaper may be composed of flint, emery, garnet, aluminum oxide, or silicon carbide. Flint is cheap but fairly ineffective over a long period; use it for removing paint. Garnet cuts better than flint and lasts longer; paper with this grit is gen-

Table 2

Retail	Old system	Industrial system	Uses
	—	600	
	—	500	
Super fine or	10/0	400	Normally not used for wood.
extra fine	9/0	360	For polishing stone, cera-
		320	mic, plastics.
Very fine	8/0	280	For polishing wood finishes
	7/0	240	between coats, usually wet
	6/0	220	sanded.
Fine	5/0	180	Fine sanding of bare
	4/0	150	wood.
	3/0	120	
		100	Soft wood sanding; shaping
Medium	2/0	100	wood.
	1/0 or 0	60	
	½	60	
Coarse	1	50	First, rough sanding.
	1½	40	Removing paint.
Very Coarse	2	36	Used for sanding floors
	2½	30	
	3	24	
Extra coarse	3½	20	Also used on floors, par-
	4	16	ticularly where floor
	4½	12	is painted.

erally useful all around. Emery is a black abrasive that has been superseded by other abrasives but lasts, it seems, simply because the newer materials are not well known.

Aluminum oxide lasts longer and cuts much better than the other three papers. Today it is regarded as the best all around sandpaper to use.

Silicon carbide is blue-black in color and is the hardest abrasive of all. It is used on metal, glass, and similarly hard substances.

SUMMARY

Various types of saws are used in carpentry work. Sawing is hard work; the craftsman should know how to saw properly and keep his saw in a sharp condition. There are many different kinds of saws, such as crosscut, rip, back, keyhole, coping, hacksaw, dovetail, drywall, and rod saw.

Files and rasps are tools used for smoothing surfaces such as metal and wood. Files are also used to sharpen saws and other cutting tools. Various shapes of files and rasps are used in carpentry work, such as flat, taper, square, round, and half-round. The Surform tool, which has a blade that looks like a cheese grater, is another useful tool in the rasp/file family.

Sandpaper is a tough paper or other material with abrasive glued to one side. It is used for smoothing and final finishing of wood surfaces. Various sizes and grades are on the market, ranging from extra fine to coarse. Sandpaper generally is purchased in 9″ × 11″ sheets, but can also be obtained in rolls and discs for various sanding machines.

REVIEW QUESTIONS

1. How many teeth per inch in the average crosscut saw?
2. What is the difference between a crosscut saw and a ripsaw?
3. What is the difference between a coping saw and a keyhole saw?
4. Explain how a hacksaw functions.
5. Why must the saw teeth be set to project beyond the surface of the blade?

CHAPTER 12

Sharpening Saws

One key to success in carpentry is maintaining sharp tools. Besides making jobs easier, it will make jobs safer. When a tool isn't sharp, you don't really know where it's going to go—and that is emphatically unsafe.

The saw is one tool that should be kept very sharp. Saws can be sharpened professionally, but many craftsmen like to do the job themselves. It's more convenient, cheaper, and a special kind of satisfaction is involved.

Besides the crosscut and ripsaw, a variety of other saws can be sharpened including the backsaw, dovetail saw, keyhole saw, and coping saw.

There are five steps in the sharpening process:

1. Jointing.
2. Shaping.
3. Setting.
4. Filing.
5. Dressing.

When sharpening handsaws, the first step is to place the saw in a suitable clamp or saw vise, as illustrated in Fig. 1. In the absence of a good saw vise, a homemade clamp may easily be made in which the saw can be supported. The saw should be held tight in the clamp so that there is no noticeable vibration. The saw is then ready to be jointed.

Jointing

Jointing is done when the teeth are uneven or incorrectly shaped or when the teeth edges are not straight. If the teeth are

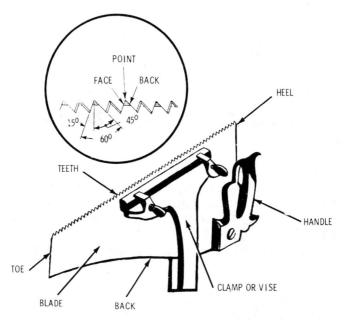

Fig. 1. The method of fastening a handsaw in a saw clamp or vise.

irregular in size and shape, jointing must precede setting and filing. To joint a saw, place it in a clamp with the handle to the right. Lay a flat file lengthwise on the teeth, and pass it lightly back and forth over the length of the blade on top of the teeth until the file touches the top of every tooth. The teeth will then be of equal height, as shown in Fig. 2. Hold the file flat; do not allow it to tip to one side or the other. The jointing tool or handsaw jointer will aid in holding the file flat.

Shaping

Shaping consists of making the teeth uniform in width, usually after the saw has been jointed. The teeth are filed with a regular handsaw file to the correct uniform size and shape. The gullets (spaces between teeth) must be of equal depth. For the crosscut saw, the front of the tooth should be filed at an angle of 15° from the vertical, while the back slope should be at an angle of 45° from the vertical, as illustrated in Fig. 3. When filing a ripsaw,

TEETH TOO HIGH

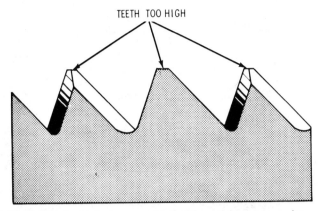

Fig. 2. The method of jointing saw teeth. Place the saw in a clamp with the handle to the right. Lay a mill file lengthwise flat on the teeth. Pass it lightly back and forth along the length of the teeth until the file touches the top of every tooth. If the teeth are extremely uneven, joint the highest teeth first, then shape the teeth that have been jointed and joint the teeth a second time. The teeth will then be the same height. Do not allow the file to tip to one side or the other; hold it flat.

PERPENDICULAR LINE

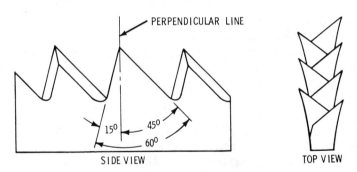

SIDE VIEW   TOP VIEW

Fig. 3. The side and tooth-edge views of a crosscut saw. The angle of a crosscut saw tooth is 60°, the same as that of a ripsaw. The angle on the front of the tooth is 15° from the perpendicular, while the back angle is 45°.

the front of the teeth are filed at an angle of 8° with the vertical, and the back slope is filed at an angle of 52° with the vertical, as shown in Fig. 4. Some good workmen, however, prefer to file ripsaws with more of an angle than this, often with the front side of the teeth almost square, or 90°. This produces a faster-cutting

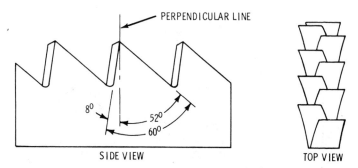

Fig. 4. The side and tooth-edge views of a typical ripsaw. The tooth of a ripsaw has an angle of 60°; that is, 8° from the perpendicular on the front and 52° on the back of the tooth.

saw, but, of course, it pushes harder, and it will grab when cutting at an angle with the grain.

When shaping teeth, disregard the bevel of the teeth, and file straight across at right angles to the blade with the file well down in the gullet. If the teeth are of unequal size, press the file against the teeth level with the largest flat tops until the center of the flat tops made by jointing is reached. Then move the file to the next gullet, and file until the rest of the flat top disappears and the tooth has been brought to a point. Do not bevel the teeth while shaping. The teeth, now shaped and of even height, are ready to be set.

Setting

After the teeth are made even and of uniform width, they must be set. Setting is a process by which the points of the teeth are bent outward by pressing with a tool known as a saw set. Setting is done only when the set is not sufficient for the saw to clear itself in the kerf. It is always necessary to set the saw after the teeth have been jointed and shaped. The teeth of a handsaw should be set before the final filing to avoid damage to the cutting edges. Whether the saw is fine or coarse, the depth of the set should not be more than one-half that of the teeth. If the set is made deeper than this, it is likely to spring, crimp, or crack the blade or break the teeth.

When setting teeth, particular care must be taken to see that

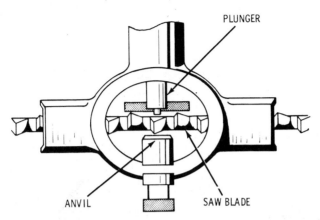

PLUNGER

ANVIL

SAW BLADE

Fig. 5. The position of the saw set on the saw for setting the teeth.

the set is regular. It must be the same width along the entire length of the blade, as well as being the same width on both sides of the blade. The saw set should be placed on the saw so that the guides are positioned over the teeth with the anvil behind the tooth to be set, as shown in Fig. 5. The anvil should be correctly set in the frame, and the handles should be pressed together. This step causes the plunger to press the tooth against the anvil and bend it to the angle of the anvil bevel. Each tooth is set individually in this manner.

Filing

Filing a saw consists of simply sharpening the cutting edges. Place the saw in a filing clamp with the handle to the left. The bottom of the gullets should not be more than ½ inch above the jaws of the clamp. If more of the blade projects, the file will chatter or screech. This dulls the file quickly. If the teeth of the saw have been shaped, pass a file over the teeth, as described in jointing, to form a small flat top. This acts as a guide for the file; it also evens the teeth.

To file a crosscutting handsaw, stand at the first position shown in Fig. 6. Begin at the point of the saw with the first tooth that is set toward you. Place the file in the gullet to the left of this tooth, and hold the handle in the right hand with the thumb and three fingers on the handle and the forefinger on top of the

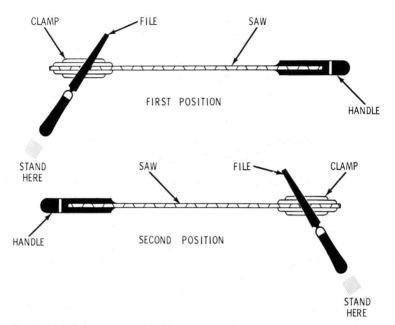

Fig. 6. Standing positions for filing a crosscut saw. The saw clamp should be moved along the blade as filing progresses.

file or handle. Hold the other end of the file with the left hand, the thumb on top and the forefinger underneath. The file may be held in the file-holder guide, as shown in Fig. 7. The guide holds the file at a fixed angle throughout the filing process while each tooth is sharpened.

Hold the file directly across the blade. Then swing the file left to the desired angle. The correct angle is approximately 65°, as shown in Fig. 8. Tilt the file so that the breast (the front side of the tooth) side of the tooth may be filed at an angle of approximately 15° with the vertical, as illustrated in Fig. 3. Keep the file level and at this angle; do not allow it to tip upward or downward. The file should cut on the push stroke and should be raised out of the gullet on the reverse stroke. It cuts the teeth on the right and left on the forward stroke.

File the teeth until half of the flat top is removed. Then lift the file, skip the next gullet to the right, and place the file in the

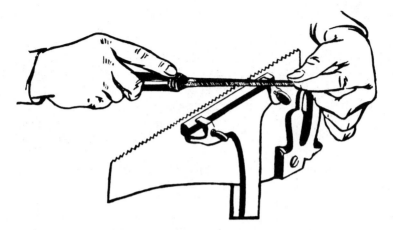

Fig. 7. The method of holding the file when filing a handsaw.

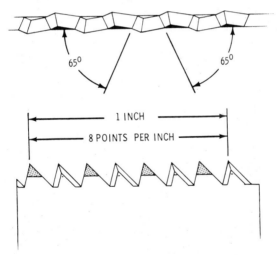

Fig. 8. The side angle at which to hold the file when filing a crosscut saw having eight points per inch.

second gullet toward the handle. If the flat top on one tooth is larger than the other, press the file harder against the larger tooth so as to cut that tooth faster. Repeat the filing operation on the two teeth which the file now touches, always being careful to keep the file at the same angle. Continue in this manner, placing

145

the file on every second gullet until the handle end of the saw is reached.

Turn the saw around in the clamp with the handle to the left. Stand in the second position, and place the file to the right of the first tooth set toward you, as shown in Fig. 6. This is the first gullet that was skipped when filing from the other side. Turn the file handle to the right until the proper angle is obtained, and file away the remaining half of the flat top on the tooth. The teeth that the file touches are now sharp. Continue the operation until the handle end of the saw is reached.

When filing a ripsaw, one change is made in the preceding operation; the teeth are filed straight across the saw at right angles to the blade. The file should be placed on the gullet so as to file the breast of the tooth at an angle of 8° with the vertical, as shown in Fig. 4. Stand in the positions shown in Fig. 9. When sharpening a ripsaw, file every other tooth from one side. Then turn the saw around, and sharpen the remaining teeth as described in the preceding paragraphs. When filing teeth, care

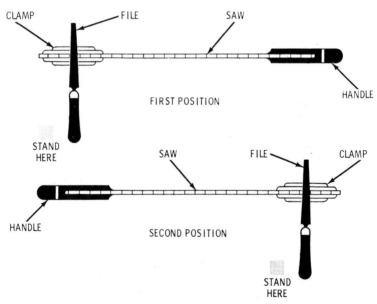

Fig. 9. *Standing positions for filing a typical ripsaw. Again, the saw clamp must be moved along the blade as filing progresses.*

must be taken in the final sharpening process to file all the teeth to the same size and height, otherwise the saw will not cut satisfactorily. Many good saw filers file ripsaws from only one side, taking care that the file is held perfectly horizontal. For the beginner, however, turning the saw is probably the most satisfactory method.

Dressing

Dressing of a saw is necessary only when there are burrs on the side of the teeth. These burrs cause the saw to work in a ragged fashion. They are removed by laying the saw on a flat surface and running an oilstone or flat file lightly over the side of the teeth.

SUMMARY

Success in woodworking depends on proper tool maintenance, which includes the tool sharpening process. There are five steps which should be considered when sharpening saws; they are jointing, shaping, setting, filing, and dressing. Filing a saw consists of simply sharpening the cutting edges.

Jointing is done when the teeth on the blade are uneven or incorrectly shaped or when the teeth edges are not straight. The high spots or high teeth are filed down and then sharpened. Shaping a saw consists of making the teeth uniform. The teeth are filed to the correct size and shape. The gullets must be of equal depth.

After the teeth are made even and uniform in width, they must be set. This is accomplished with the use of a tool known as a saw set. It is necessary to set a saw when the teeth have been jointed and shaped. The saw must be placed in the clamp properly and held solid.

REVIEW QUESTIONS

1. What are the five steps used in sharpening a saw?
2. What is the heel on a handsaw?
3. What is jointing a saw?
4. How are jointing and shaping accomplished?
5. Explain the process of dressing a saw.

CHAPTER 13

Sharp-Edged Cutting Tools

There is a whole array of sharp-edged cutting tools that are considered as hand guided—such as chisels, drawknives, and the like—as opposed to striking tools such as hatchets and axes, and self-guided tools such as planes. While some cutting tools are considered more essential than others (for example, chisels are used more often than drawknives) all have their place. And even if some tools are not used that much—or at all—you should know about them so that you can use them should the need arise.

CHISELS

In carpentry, the chisel is a very important tool. It is, however, one of the most abused tools, because it is often used for prying open things and even as a screwdriver, although it is designed solely for cutting wood surfaces. A chisel consists of a flat, thick piece of steel with one end ground at an acute bevel to form a cutting edge and the other end provided with a handle, as shown in Fig. 1.

Chisels may be classed:
1. With respect to duty or service, as:
   a. Paring.
   b. Firmer.
   c. Gouge.

2. With respect to the length of the blade, as:
   a. Butt.
   b. Pocket.
   c. Mill.
3. With respect to the edges of the blade, as:
   a. Plain.
   b. Bevel.
4. With respect to the method of attaching the handle, as:
   a. Tang.
   b. Socket.
5. With respect to the shape of the blade, as:
   a. Flat.
   b. Round (gouge).
   c. L (corner).

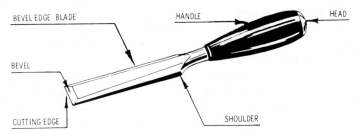

BEVEL EDGE BLADE   HANDLE   HEAD

BEVEL

CUTTING EDGE   SHOULDER

Fig. 1. A typical general-purpose wood chisel with a handle of hard plastic, which is good at resisting breakage.

Paring Chisel

This is a light-duty tool for shaping and preparing relatively long planed surfaces, especially in the direction of the grain of the wood. The paring chisel (Fig. 2) is manipulated by a steady sustained pressure of the hand; it should not be driven by the blows of a hammer or other similar tool.

Firmer Chisel

The term "firmer" implies a more substantial tool than the paring chisel, and is adapted to medium-duty work. The firmer chisel (Fig. 2) is a tool for general work and may be used either for paring or light mortising; it is driven by hand pressure in paring and by blows from a mallet in mortising.

NOTE: A hammer or other tool should not be used to drive a wood-handled chisel—use only a mallet. Wood to wood in driving is the only satisfactory method for driving chisels.

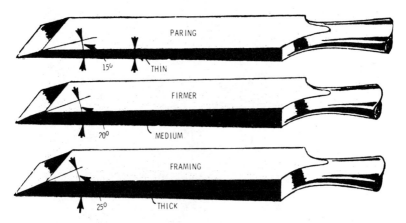

Fig. 2. *Paring chisel is for light duty; firmer for medium duty. Chisels differ in the blade thickness. Paring is thin, firmer, thicker.*

Gouge

This is a chisel with a hollow-shaped blade for scooping or cutting round holes. There are two kinds of gouge chisels, the outside bevel and the inside bevel, as shown in Fig. 4. The outside bevel is the more common of the two.

Tang and Socket Chisels

According to the method by which the blade and handle are joined, chisels are called tang or socket. The difference in the two types is shown in Fig. 5. The tang chisel has a projecting part, or tang, on the end of the blade which is inserted into a hole in the handle. The reverse method is employed in the socket chisel; that is, the end of the handle is inserted into a socket on the end of the blade. The term "socket firmer," as applied to a firmer chisel having a socket end, does not mean (as generally supposed by some craftsmen) "hit it firmer," although that is what actually happens in operation; the blows tend to drive the handle firmer into the tapered socket.

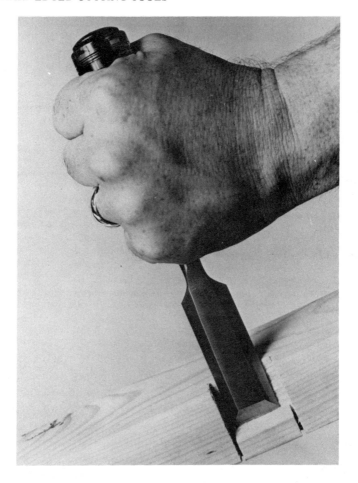

Fig. 3. The butt chisel has a relatively short blade. Like all chisels, it is designed to be tapped on top of handle, and it cuts with the flat side down. Courtesy of Stanley.

Butt, Pocket, and Mill Chisels

This classification relates simply to the relative lengths of the blades. The regular lengths, shown in Fig. 6, are approximately as follows: butt, 2½ to 3¼ inches, pocket 5 to 6 inches, and mill, 8 to 10 inches.

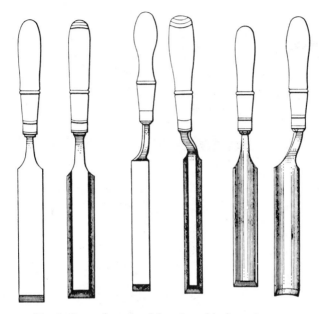

Fig. 4. Several types of framing chisels and gouges.

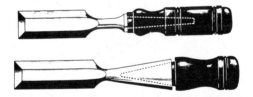

Fig. 5. Typical tang (top) and butt socket chisels. The terms "tang" and "socket" are derived from the fact that the shank of the tang chisel has a point which is fitted into the handle. This point is called a tang, hence the name "tang chisel." In the socket chisel, the shank of the chisel is made like a cup, or socket, with a handle fitted into it; thus the chisel is called a "socket chisel."

How To Select Chisels

A chisel should be absolutely flat on the back (the sides should not be beveled). An inferior chisel is ground off on the back near the cutting edge, with the result that, in use, it tends to follow the grain of the wood, splitting it off unevenly, since the user cannot properly control his tool. The flat back allows the chisel

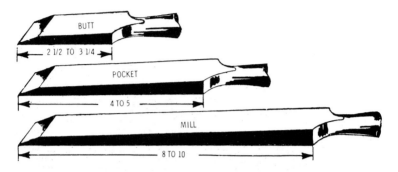

Fig. 6. Wood chisels classified with respect to the length of the blade. The blade width may vary, depending on the type of work to be performed. The lengths given in the illustration are in inches.

to take off the finest shaving, and where a thick cut is desired, it will not strike too deep. This is an important point to be looked for in good chisels.

The best chisels are made of selected steel with the blade widening slightly toward the cutting edge. The blades are oil-tempered and carefully tested. The ferrule and blade of the socket chisel are so carefully welded together that they practically form a single piece. Socket chisels are preferred to the tang type by most carpenters, because they are stronger and the handles are less apt to split. Beveled edges are preferable to plain blades, because they tend to drive the tool forward and also have greater clearance.

The chisels commonly carried on construction jobs these days are quite short—only 3- or 3½-inch beveled blades and metal-capped handles of a tough plastic that will take an amazing amount of abuse. The handles are cast and molded in place and are intended to be driven with a hammer. Such chisels are too short for a good job of paring or deep mortising but are service-able for general heavy-duty use and for door-butt and similar mortising.

One excellent handle for socket-firmer chisels is sometimes, but rarely, available. It is built up of sole-leather rings around a steel rod core. These handles are almost indestructible and may be driven with a hammer.

The butt chisel, because of its short blade, is adapted for close accurate work where not much power is required. It is particularly suited for putting on small hardware, which does not necessarily require the use of a hammer. The butt chisel may be used almost like a jackknife with the hand placed well down on the blade toward the cutting edge. The short blade and handle make it convenient for carrying in the pocket. Chisels are usually ground sharp and hand honed and are ready for use when sold.

How To Care For and Use Chisels

In order to do satisfactory work with chisels, the following instructions should be carefully noted and followed:

1. Do not drive the chisel too deep into the work; this requires extra pressure to dig out the chips.
2. Do not use a firmer chisel for mortising heavy timber.
3. Keep the tool bright and sharp at all times.
4. Protect the cutting edge when not in use.
5. Never use a chisel to open boxes, to cut metal, or as a screwdriver or putty knife or prying tool.

Perhaps it should be emphasized that the chisel (and the screwdriver) takes more direct assaults from craftsmen than any other tool, not only in terms of care but in terms of faulty use. For these reasons the care practices suggested above should be followed closely, and the chisel should be carefully used. Fig. 7 illustrates the right and wrong ways to use the paring chisel, while Fig. 8 shows the proper use of a firmer chisel. Fig. 9 shows how to use chisels for various types of work.

How To Sharpen Chisels

When honing a chisel, use a good grade of oilstone. Pour a few drops of machine oil on the stone; if you have no machine oil, lard can be used. The best results are obtained by using a carborundum stone. The carborundum cuts faster than most other abrasives, but the edge will not be as smooth and keen as when a natural oilstone is used.

Hold the chisel in the right hand, and grasp the edges of the

155

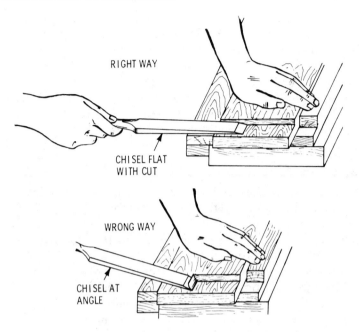

Fig. 7. The right and wrong ways to use a paring chisel. The flat side of the chisel should face the cut and be parallel with it.

stone with the fingers of the left hand to keep it from slipping. A better method is to place the stone on a bench and block it so it cannot move; both hands will then be free to use for honing. In this case, grasp the chisel in the right hand where the shoulder joins the socket. Place the middle and forefinger on the blade near the cutting edge. Rub the chisel on the oilstone away from you; be careful to keep the original bevel.

Never sharpen the chisel on the back or flat side; this should be kept perfectly flat. For paring, the taper should be long and thin—approximately 15°, as shown in Fig. 2. The longer the bevel on the cutting edge, the easier the chisel will work, and the easier it is to hone. A firmer chisel should be ground at an angle of not less than 20°; an angle of 25° is recommended for a framing chisel. When honing a chisel, the taper should be carefully maintained, and, unless the back is kept flat, it will be impossible to work to a straight line. Bevel-edge chisels are more easily

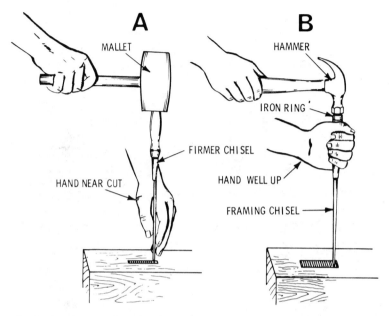

Fig. 8. Mortising with a chisel; using a firmer chisel for light mortise work (note the position of the hand and the type of driving tool used).

sharpened than the plain-edge type because there is not as much steel to be removed.

If the chisel is badly nicked, it will have to be ground before honing. Not many quality chisels can be filed. Do not overheat or damage the temper of the chisel, and be sure to keep the original taper of the bevel. After grinding, hone the chisel on an oilstone as detailed earlier.

DRAWKNIFE

This tool consists of a large sharp-edged blade with a handle at each end, usually at right angles to the blade, as shown in Fig. 10. It is used for trimming wood by drawing the blade toward the user. When the blade is sharp, and some degree of force is applied, it does its work quickly and efficiently.

The tool was formerly used to a great extent for the rapid

A. The use of a firmer chisel for light mortise work.   **B. The use of a paring chisel.**

C. The method of cutting a concave curved corner.

Fig. 9. Using chisels for various types of work.

reduction of stock to an approximate size, an operation that is now to a greater extent performed by sawing or planing machines. The drawknife is, however, quite effective on narrow surfaces that must be considerably reduced; the work is first trimmed with an adze or hand axe. Drawknives are made with cross sections of various shapes, thereby adapting them for a variety of jobs, as illustrated in Fig. 11.

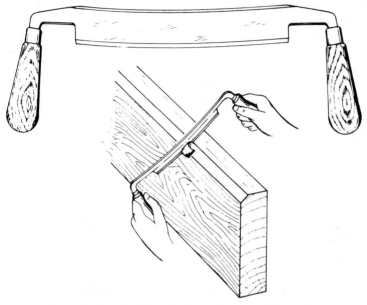

Fig. 10. A typical drawknife and the way it's used.

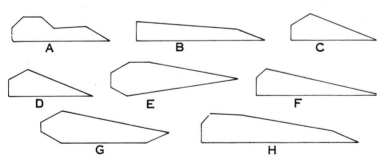

Fig. 11. Cross sections of various drawknife blades; A. Carpenters' razor blade; B. Carpenters' light blade; C. Carriage makers' narrow blade; D. Coach makers' razor blade; E. Wagon makers' heavy blade; F. Shingle shave, heavy blade; G. Saddle tree shave, heavy blade; H. Spar shave, heavy blade.

SUMMARY

Poor results in woodworking often result from using dull cutting tools; that is, neglecting to keep tools sharp. Not only

159

should the cutting edge be whetted or honed as soon as any sign of dullness is observed, but the tools should always be kept perfectly clean and free from rust. Sharp-edged cutting tools may be divided into several classes, such as chisels, planes, drawknives, hatchets, and axes.

When honing a chisel, use a good grade of oilstone. Pour a few drops of machine oil on the stone and rub the chisel away from you, being careful to keep the original bevel. Never sharpen a chisel on the back or flat side—this should be kept perfectly flat. The best chisels are made of selected steel with the blade widening slightly toward the cutting edge.

REVIEW QUESTIONS

1. What is the difference between a tang chisel and a socket chisel?
2. What is a paring chisel?
3. What is a drawknife?
4. What are the three basic blade lengths for chisels?
5. What is a corner chisel?

CHAPTER 14

Axes and Hatchets

At one time the axe and the hatchet (really a one-handed axe) were of great importance in building, but today's structures are constructed more with saws and power tools than with such relatively crude tools. Still, the axe and especially the hatchet are used by some people in building, to trim a piece of framing—whittle it down to size—for sharpening stakes, and to serve as a hammer. And specific kinds of hatchets are useful, even crucial, to the work of craftsmen who install certain kinds of siding and roofing. At any rate, the axe and the hatchet are tools the craftsman should have a knowledge of and know how to use.

BROAD HATCHET OR HAND AXE

This is simply a large hatchet with a broad cutting edge. Ordinarily it is grasped with the right hand at a distance of approximately one-third from the end of the handle, but the position of the hand will be regulated to a great measure by the material to be cut, that is, by the intensity of the blow, as illustrated in Fig. 3. Thus, to deliver a heavy blow, the handle is grasped close to the end, and for a light blow, it is held nearer to the head of the axe.

AXE

This tool is similar to the hand axe or hatchet but is of a larger size and has a longer handle and is intended for heavy cutting

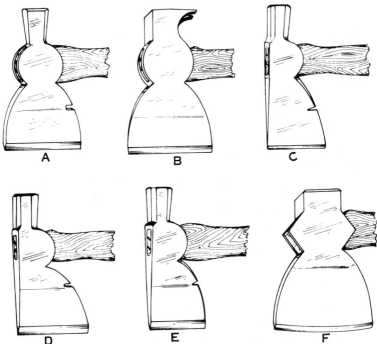

Fig. 1. Various types of hatchets. A. Shingling; B. Claw; C. Barrel; D. Half; E. Lath; F. Broad (hand axe).

using both hands. A typical axe is shown in Fig. 5 at the bottom of the illustration.

ADZE

An adze is a form of hatchet in which the edge of the blade is at right angles to the handle. The blade is curved or arched toward the end of the handle, thus permitting an advantageous stroke of the tool while the operator is standing over the work. The edge is beveled on the inside only, and the handle can be removed when it is necessary to grind the tool.

When sharpening an adze, the tool should be traversed across the face of the stone; hold it at the proper bevel angle until all the nicks have been taken out. Then to secure a keen edge, rub it with a slipstone. It is important when sharpening an adze to bevel only on the inside.

162

Fig. 2. A hatchet such as the one shown is being used to establish length of shingle to the weather—how much is exposed. The gauge pin in the hatchet is adjustable to "weather" length desired. Striking face is beveled and tempered for driving nails. Courtesy of Vaughn & Bushnell.

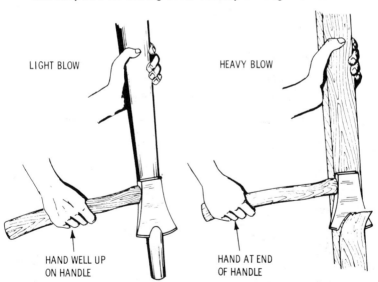

LIGHT BLOW

HEAVY BLOW

HAND WELL UP
ON HANDLE

HAND AT END
OF HANDLE

Fig. 3. Grasp the handle of the hand axe approximately halfway between the ends to strike a light blow and at the end of the handle to obtain the necessary swing for a heavy blow.

163

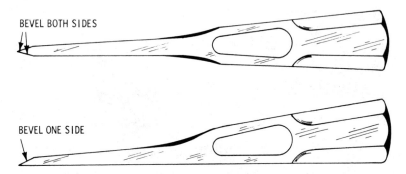

Fig. 4. When sharpening a hatchet or axe, bevel both sides of the cutting edge for general use; for hewing a line, bevel only one side.

Fig. 5. Typical hatchets and an axe.

SUMMARY

A hatchet can be used as a hammer, for sharpening stakes, cutting timber to a rough size, or splitting wood. The hatchet is also used in roof shingling.

There are various types of hatchets, such as shingling, claw, barrel, half, and lath. The cutting edge must, at all times, be free of nicks in order to be a useful tool.

REVIEW QUESTIONS

1. What is the difference between a hatchet and an axe?
2. What is an adze?
3. In what type of work would a hatchet be used?
4. How are hatchets used in framing work?

CHAPTER 15

Smooth Facing Tools

The tools under this classification are those sharp-edged cutting tools in which the cutting edge is guided by the contact of the tool with the work instead of being guided by hand. For example, consider a plane as distinguished from a chisel. The plane, being positively guided, gives a smooth cut in contrast with the rough cut obtained by the hand-guided chisel—hence the term "smooth facing tools." These tools are essentially chisels set in appropriate frames, so that the contact of the frame with the work during the movement of the tool will give a positive guide to the cutting edge, thus resulting in a smooth cut.

THE SPOKESHAVE

This tool resembles a modified drawknife, whose blade is set in a boxlike frame which forms a positive guide. The blade is adjustable like a plane to govern the thickness of the cut. Spokeshaves may be made of wood or metal, and they have obtained their name from the fact that they were once used in the making of wagon spokes. They are very useful in smoothing curved edges and to round irregular surfaces, as shown in Fig. 1. Spokeshaves are made with cutters of various shapes—straight, hollow, round, angular, etc. A typical spokeshave is shown in Fig. 2.

The flat-bottom spokeshave is used on convex and concave surfaces where the curves have a long sweep; it is also used to chamfer or round edges. The hollow-bottom, or concave-bottom, spokeshave is used for rounding edges that have small

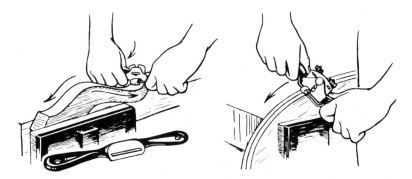

Fig. 1. The method of using the spokeshave. The spokeshave cuts when pushed away from the user, as indicated by the arrow. Be careful to work in the direction in which the tool cuts without tearing the grain. The spokeshave is also used to chamfer and cut edges.

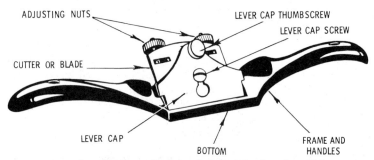

Fig. 2. A typical spokeshave. This is a lightweight, handy tool for use on concave, curved edges that have large or small sweeps. The wing-nut adjustment is on the cutter cap.

convex sweeps, and the convex-bottom spokeshave is employed to cut concave curved edges that have small sweeps.

Spokeshave cutters may be sharpened by removing the blade from the stock and rubbing it on the inside with a flat slip of oilstone; lightly rub the outside of the blade on an ordinary oilstone. To more firmly hold the small blade, place it into a saw kerf made across the end of a small, flat piece of wood, with the edge of the blade projecting beyond the wood. The piece of wood should be beveled to allow the blade to lie on the stone at the proper angle. It may then be sharpened like a plane iron.

PLANES

A plane is a tool for smoothing boards or other surfaces of wood. It consists of a stock (usually made of wood or iron or a combination of both) from the underside, or face, of which slightly projects the cutting edge. The cutting edge, which inclines backward, is called the plane iron. An aperture in the front provides for the escape of the shavings, which are produced when the tool is in action.

The plane is essentially a finishing tool, and, while it is adapted for use in bringing wood surfaces to the desired thickness, it will produce this result only gradually as compared to a chisel or hatchet. For this reason, it is normally the last tool to be used in finishing a wood surface.

There are many planes (some of which are shown in Fig. 3) to meet varied requirements. Useful planes include:

1. Jack plane.
2. Fore plane.
3. Jointer plane.
4. Smoothing plane.
5. Block plane.
6. Molding plane.
7. Rabbet plane.
8. Grooving plane.
9. Router plane.

Jack Plane

This plane is intended for heavy, rough work. It is generally the first plane used in preparing the wood; its purpose is to remove irregularities left by the saw and to produce a fairly smooth surface. The jack plane is long and heavy enough to make it a powerful tool, so that it will remove a considerable chip with each cut. The cutting edge of the plane iron is ground slightly rounded; this form is best adapted for roughing. If properly sharpened, the jack plane may be used as a smoothing plane, or as a jointer on small work, because it is capable of doing just as good work.

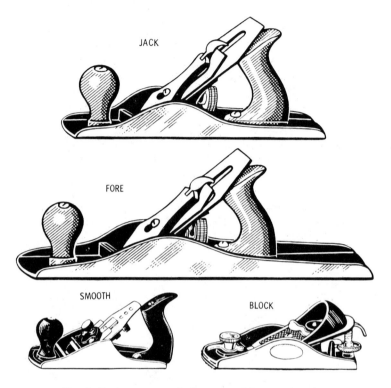

JACK

FORE

SMOOTH

BLOCK

Fig. 3. Various types of planes used by woodworkers.

Fore Plane

This plane is designed for the same purpose as the jointer plane, that is, to straighten and smooth the rather rough and irregular cut of the jack plane. Since the fore plane is shorter than the jointer (usually 18 inches in length), it is easier to handle, especially for a journeyman carpenter, and it may also be used as a jack plane. If a carpenter does not have both a jack plane and a jointer, he can make a fore plane serve for both, although it will not give as good service as either of the other two in the work for which they are adapted. The plane iron of the fore plane is sharpened to a straight line and is set for a finer cut than that of the jack plane.

Jointer Plane

The great length and weight of these planes keep the cutter from tearing the wood, and, with the cutter set for a fine cut, it is the plane to use for obtaining the smoothest finishes. These planes will true up better than other types of planes.

In this country, the word "jointer" is applied to planes that range in size from 22 to 30 inches. The length of the plane determines the straightness of the cut. Thus, a smoothing plane, because of its short length, will follow the irregularities of an uneven surface, taking its shavings without interruption, whereas a fore or jointer plane similarly used will first touch only the high spots, progressively lengthening the cuts until, on reaching the lowest spots, a continuous shaving will be taken. The final cut will approach a true surface depending on the length of the tool and the length of the irregularities or undulations that were there to begin with. The cutting edge of a jointer plane is ground straight and is set for a fine cut.

Smoothing Plane

The small length of this plane (usually about 8 inches) adapts it for finishing uneven surfaces: because of its small size, it will find its way into minor depressions of the wood without taking off much material. In this respect, the smoothing plane differs from the jointer plane; although both are finishing planes, the jointer plane is used for finer work. A typical smoothing plane is shown in Fig. 4.

Block Plane

This type of plane, shown in Fig. 5, is the smallest plane made (length 4 to 7 inches). It was designed to meet the demand for a plane that may easily be held in one hand while planing across the grain. The block plane is used almost exclusively for planing across the grain; therefore, no cap iron is necessary to break the shavings, since they are only chips.

The bevel of the plane iron is turned up instead of down. Because of its size, the block plane is usually operated with one hand, with the work held by the other hand. Therefore, as distinguished from this method of using, other planes are called bench planes. The angle of the plane iron for block planes is

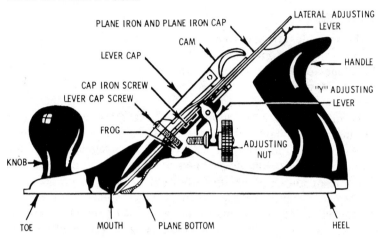

Fig. 4. A typical smoothing plane and its component parts.

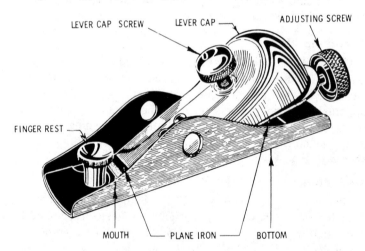

Fig. 5. A typical block plane and its component parts.

much smaller than that for bench planes. This angle is 20° for softwoods and 12° for hardwoods; planes with the iron set at 12° are called "low angle" block planes.

Rabbet Plane

In this type of plane, as shown in Fig. 6, the plane iron projects slightly from the side as well as from the bottom of the

172

plane. There are various forms of rabbet planes available, each suitable for different types of cuts. With a tool of this type, the edge of a board can be cut so as to leave a rabbet or "sinking" (like a step) along its length to fit over and into a similar indentation cut in the edge of another board. Rabbet planes are adapted to cut with or across the grain according to the setting of the iron.

Fig. 6. A typical rabbet plane.

Surform

The Surform from Stanley Tools is also in the plane family. Like other tools in this line, it has a cheese-grater-like cutting surface and makes short work of wood, soft metals, and other materials. The tool is pushed along like any plane to do its cutting.

Grooving Plane

This plane (sometimes called a trenching plane) is used for cutting grooves across the grain. It has a rabbeted sole; the cutters are in the tongue portion, which is usually ½ inch deep and varies from ¼ to 1⅛ inches. A screw stop adjusts the depth of the cut, and a double-toothed cutter separates the fibers in front of the iron. A typical grooving plane is shown in Fig. 8.

Router

Planes of this type are used for surfacing the bottoms of grooves or the like which are parallel with the general surface of the wood. The closed-throat type is the ordinary form of router;

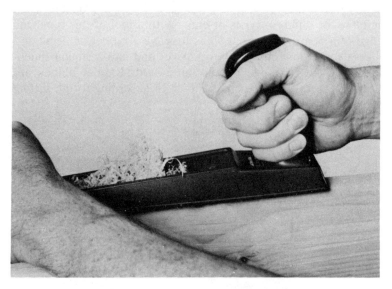

Fig. 7. A Surform plane from Stanley. It removes a large amount of material quickly. Courtesy of Stanley.

Fig. 8. A typical grooving, or plow, plane, especially designed for weatherstrip grooving.

the open-throat type is an improved design, giving more freedom for chips and a better view of the work and cutter. Both types are shown in Fig. 9. The open-throat router has an attachment for regulating the thickness of the chip and a second attachment for closing the throat for use on narrow surfaces. The bottoms of both styles are designed so that an extra wooden bottom of any size desired can be screwed on, thereby enabling the user to rout large openings.

Planes Are Expensive

It should be noted that planes can be particularly expensive. Indeed, a number of them cost far more than a router, a portable

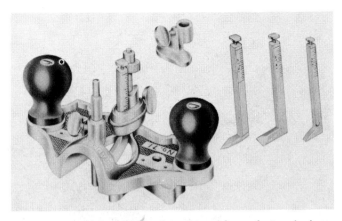

Fig. 9. Typical router planes. This plane is used for surfacing the bottom of grooves or other depressions parallel to the work.

power tool that can do so many of the jobs the plane does. Of course, the satisfaction of working with a hand tool like a plane is great, but you may want to consider the cost.

Plane Irons or Cutters

The so-called plane iron that does the cutting is similar to a chisel but differs in that its sides are parallel and the thickness is less than that of the chisel blade.

Plane irons are classed:

1. With respect to thickness:
   - *a.* Heavy.
   - *b.* Thin.
2. With respect to the shape of the cutting edge:
   - *a.* Curved.
   - *b.* Straight (square)
3. With respect to provisions for breaking the chips:
   - *a.* Single (Fig. 11).
   - *b.* Double (Fig.12).

Heavy plane irons are usually No. 12 gauge, whereas the medium or thin plane irons are usually No. 14 gauge. The heavy plane iron offsets the tendency found in spring-cap planes to vibrate; the additional weight helps avoid chattering.

Fig. 10. The iron on a plane can be easily removed for sharpening or manipulated for setting. Courtesy of Stanley.

The thin plane iron is normally satisfactory when the plane is properly constructed; that is, firm support is given the cutter over a considerable portion of its length. It also has the slight advantage of requiring less grinding.

For the first, or roughing, cut with the jack plane, the cutting edge is ground slightly curved (convex), as shown in Fig. 13, because, since it is used for heavy work, it removes thick shavings; if the cutter were ground straight, the plane would cut a

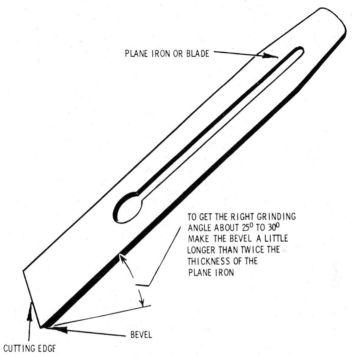

PLANE IRON OR BLADE

TO GET THE RIGHT GRINDING
ANGLE ABOUT 25° TO 30°
MAKE THE BEVEL A LITTLE
LONGER THAN TWICE THE
THICKNESS OF THE
PLANE IRON

BEVEL

CUTTING EDGE

Fig. 11. A single plane iron.

rectangular channel from which the wood must be torn as well as cut, as shown in Fig. 14A. Moreover, such a shaving would probably stick fast in the throat of the plane or require undue force to push the plane. Compare this with the shaving taken from the fully curved cutting edge of the jack plane, shown in Fig. 14B.

When a full set of planes is available, the fore plane should have some curvature to the cutting edge. In this case, the process of transforming the grooved surface produced by the jack plane to a flat surface is accomplished in three operations, using the jack, fore, and jointer plane, as shown in Fig. 15. The cutting edge of the jointer and smoothing plane irons are made straight with rounded corners, as shown in Fig. 13. Because this type of plane iron makes an extremely fine cut, the groove caused by

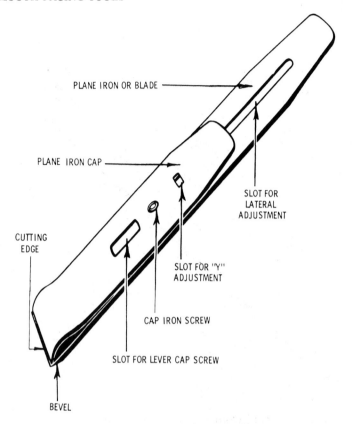

PLANE IRON OR BLADE

PLANE IRON CAP

SLOT FOR LATERAL ADJUSTMENT

CUTTING EDGE

SLOT FOR "Y" ADJUSTMENT

CAP IRON SCREW

SLOT FOR LEVER CAP SCREW

BEVEL

Fig. 12 A double plane iron.

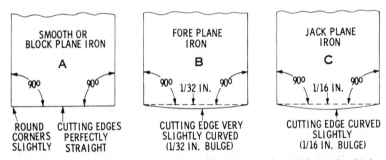

SMOOTH OR BLOCK PLANE IRON

A

90° 90°

ROUND CORNERS SLIGHTLY

CUTTING EDGES PERFECTLY STRAIGHT

FORE PLANE IRON

B

90° 1/32 IN. 90°

CUTTING EDGE VERY SLIGHTLY CURVED (1/32 IN. BULGE)

JACK PLANE IRON

C

90° 1/16 IN. 90°

CUTTING EDGE CURVED SLIGHTLY (1/16 IN. BULGE)

Fig. 13. Cutting edges for common plane irons. These edges should be straight on smooth and block plane irons and just slightly curved on jack and fore plane irons.

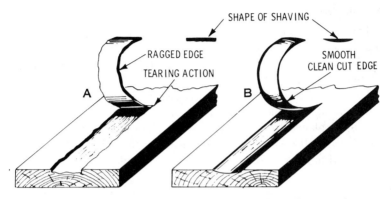

Fig. 14. The actions of a jack plane with a straight and a curved cutter.

the removal of so delicate a shaving is sufficiently blended with the general work by the rounded corners of the iron.

Bevel of the Cutting Edge

Many of the complaints concerning poorly cutting plane irons are due to improper plane-iron grinding. The bevel should always be ground at an angle of 25° to 30°; this means it must be twice as long as the cutter is thick. If the bevel is too long, the plane will jump and chatter; if it is too short, it will not cut. It is a good rule, perhaps, to have a long thin bevel for softwood and a 25° bevel for the hardwoods, although cross-grained timber requires a short bevel.

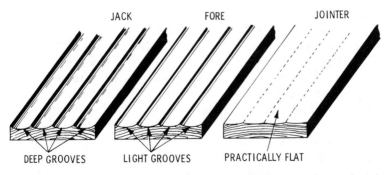

Fig. 15. The appearance of a board surface after having been planed with jack, fore, and jointer planes.

Double Irons

The term "double iron" means a plane iron equipped with a supplementary iron called a "cap." The object of the cap is to break the shaving as soon as possible after it is cut. The action of the cap is shown in Fig. 16B.

The cap is attached to the cutting iron by tightening a screw which passes through a slot in the cutter. The distance at which the cap is placed from the edge of the plane iron varies with the thickness of the shaving; allow $^{1}/_{32}$ inch for a smooth or fore plane and approximately $^{1}/_{8}$ inch for a jack plane.

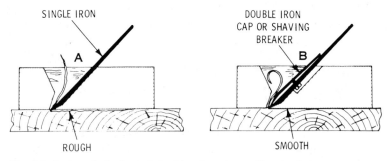

SINGLE IRON   DOUBLE IRON CAP OR SHAVING BREAKER

A   B

ROUGH   SMOOTH

Fig. 16. Action of single and double plane irons. The single iron cuts satisfactorily only when the grain is favorable, but when the grain varies from the line of cut, the shavings run up the iron, thereby leaving a rather rough surface.

Plane Mouth

This is the rectangular opening in the face of the plane through which the cutter projects and, in operation, through which the shavings pass. The width of the mouth has an important bearing on the proper working of the plane. That portion of the plane face in front of the mouth prevents the wood from rising in the form of a shaving before it reaches the mouth. If there was no face in front of the cutter, as in the case of a bull-nose plane, there would be nothing to hold down the wood in advance of the cutter, and the shaving would not be broken. In obstinate grain, the work will be rougher, and a splitting instead of a cutting action may result, as illustrated in Fig. 17. Accordingly, the wider the mouth, the less frequently the shaving will be broken and, in tough grain, the rougher the work.

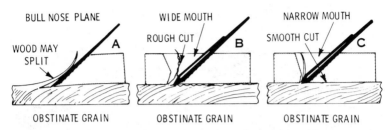

Fig. 17. Influence of mouth width. *The bull-nose plane may be regarded as an ordinary plane with a mouth of infinite width; since there is nothing in front of the cutter to hold down the wood, a splitting action is possible with obstinate grain. B and C show the results obtained with wide- and narrow-mouth planes.*

How To Use a Plane

In order to obtain satisfactory results with planes, it is necessary to know not only the proper method of handling the tool in planing but also how to put it into good working condition. The user must know:

1. How to sharpen the cutter.
2. How to adjust the cutter.
3. How to plane.

Sharpening Planes — This involves two operations, grinding and whetting. When grinding, the cutter must be ground perfectly square; that is, the cutting edge must be at right angles to the side. Enough metal must be removed to remove all nicks in the cutting edge. Before grinding, loosen the cap and set it back approximately ⅛ inch from the edge; it will serve as a guide by which to square the edge.

The cutter should be held firmly on the grinding wheel at the proper angle and should be moved continually from side to side to prevent wearing the stone out of true. Grind on the bevel side only. The bevel angle should be approximately 30°. This angle is attained when the length of the bevel is twice the thickness of the cutter. Fig. 18 shows the proper position of the cutter on the grinding wheel. The edge should be ground to one of the forms shown in Fig. 13, depending on the type of plane and the requirements of the work to be done.

After grinding, the cutter will have a wire edge; that is, the

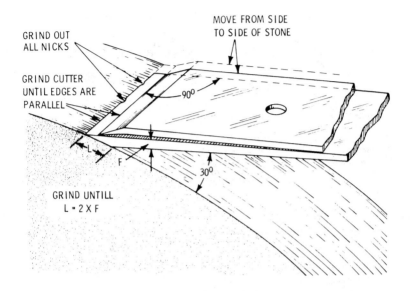

GRIND OUT ALL NICKS

MOVE FROM SIDE TO SIDE OF STONE

GRIND CUTTER UNTIL EDGES ARE PARALLEL

90°

L

F

30°

GRIND UNTILL L = 2 X F

Fig. 18. The proper position of the plane iron on the grinding wheel. Note the conditions which must be fulfilled to grind properly.

coarse grit of the grinder will always leave the edge comparatively coarse or rough, and the edge will not be as keen as it should be to cut smoothly. This wire edge is removed by the aid of an oilstone, as shown in Fig. 19.

In the case of a double iron, the cap should be kept with a fine but not a cutting edge. The cap must be made to fit the face of the cutter accurately; if it does not fit precisely, the plane will quickly "choke" with shavings because of the shavings driven between the two irons. This is an extremely important point, and it should be noted that even a minute opening between the plane irons will allow the shavings to drive in and choke the plane.

To make the sharpening job easier there is a honing guide available. The plane iron can be inserted in the device and set for the proper angle, then the blade can be drawn back across the stone with no errors possible. One such guide is good for blades up to 2⅜ inches wide. The device can also be used for sharpening spokeshave and woodchisel blades.

Adjusting the Cutter—After sharpening the cap of a double

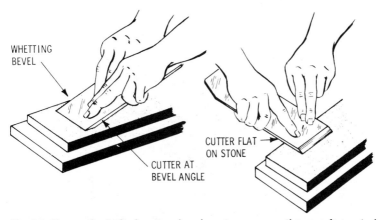

WHETTING BEVEL

CUTTER AT BEVEL ANGLE

CUTTER FLAT ON STONE

Fig. 19. The method of whetting the plane iron on an oilstone after grind-
ing. Grasp the plane iron firmly in the right hand with the palm down,
pressing down with the left hand near the cutting end to provide rubbing
pressure. Rub back and forth along the length of the stone. After whet-
ting the bevel side, turn the plane iron over and hold it perfectly flat on
the stone; give it two or three strokes to remove any wire edge.

iron, position the screw in the slot; tighten the screw lightly on the cap to within ¼ inch of the cutting edge, then tighten the screw. Finish the setting by driving the cap up to its final position, tapping lightly on the setscrew.

The "set" of the iron is the amount of cutter face exposed below the edge of the cap. The plane iron is said to be set coarse or fine according to the amount of cutter face exposed. The set, therefore, regulates the thickness of the shavings and is varied according to the nature and kind of wood to be planed. For softwoods, the set should be: ½ inch for the jack plane, $1/16$ inch for the jointer plane, and $1/32$ inch for the smoothing plane. If the wood is hard or cross-grained, allow for approximately one-half of these settings.

How To Plane—Satisfactory results in the use of a plane depend largely on the plane being in perfect condition and properly adjusted with respect to set and depth of cut to suit the kind of wood to be planed. The first thing to learn is the correct way of holding the plane. The plane should not drop over the end of the board at either end of the stroke. Before planing, examine the board with respect to the grain, and turn the material to take

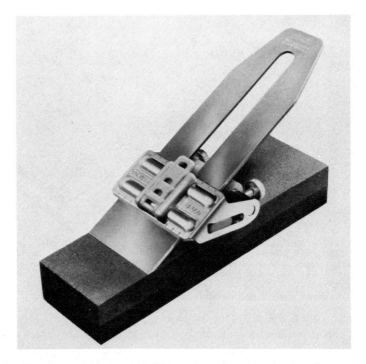

Fig. 20. A honing guide makes sharpening plane irons easier. Courtesy of Stanley.

advantage of the grain. On the return stroke, lift the back of the plane slightly so that the cutter does not rub against the wood, thus preventing the cutter from being quickly dulled.

When planing a narrow surface, let the fingers project below the plane, and use the side of the board as a guide to keep the plane on the work. If the plane chokes with shavings, look for and repair the cause instead of just removing them. Remove the iron and carefully examine the edge of the cap. This must be a perfect fit or there will be continual trouble.

To plane a long surface, such as a long board, begin at the right-hand end. Take a few strokes, then step forward and take the same number of strokes, progressing this way until the entire surface is passed over. To preserve the face of the plane, apply a few drops of oil occasionally to the cutter.

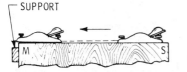
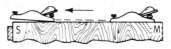

Fig. 21. The method of using the block plane across the grain. The cut must not be taken across the entire length of the board to prevent the board from splitting. Lift the plane before the cutter runs off the edge. Take several strokes with the board in position MS; then reverse the board to position SM and continue planing.

When cutting across the grain with a block plane, the cut should not be taken entirely across, but the plane should be lifted before the cutter reaches the edge of the board; if this precaution is not taken, the wood will split at the edge. After taking a few strokes, reverse the board and continue as directed, as shown in Fig. 21.

Take care when planing the edges of a piece to avoid splitting off ends. It's safest to work from the edge in toward the center of the work. If you are planing all four edges of a board, it may be practical to plane in a clockwise or counterclockwise fashion so that if the planing chips adjacent edges, you can shave these off on subsequent passes.

SCRAPERS

The term "scraper" usually signifies a piece of steel plate of approximately the thickness and hardness of a saw. There are several types of scrapers:

1. Unmounted.
2. Handle scraper.
3. Scraper plane.

The unmounted scraper is simply a rectangular steel blade, whose cutting edges are formed by a surface which is at right angles to the sides. Quicker cutting is secured with the cutting edge at a more acute angle, but more labor is required to keep it sharp. The cutting edge is sharpened by filing or grinding. For smooth work, the roughness of the edge may be removed by an oilstone, but the rougher edge will cut faster.

185

Fig. 22. When planing near an edge, work from the edge in toward the center. Courtesy of The American Plywood Assn.

When cutting, the scraper is inclined slightly forward and is more conveniently held when provided with a handle or mounted like a plane. A typical scraper is shown in Fig. 23. The method of using a hand scraper is illustrated in Fig. 24.

SUMMARY

The woodworking plane consists essentially of a smooth-soled stock of wood or iron, from the underside (or face) of which projects the steel cutting edge. The plane iron is that part of the cutting edge or knife. A section in the front provides an escape for the shavings.

There are various types of planes, such as jack, jointer, smoothing, block, molding, rabbet, grooving, and router. The

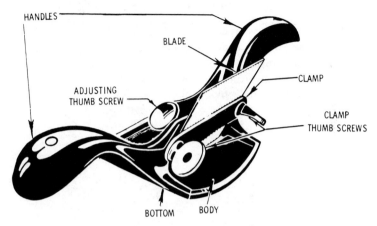

HANDLES

BLADE

ADJUSTING
THUMB SCREW

CLAMP

CLAMP
THUMB SCREWS

BOTTOM   BODY

Fig. 23. A typical cabinet scraper. In operation, the blade springs backward, thereby opening the mouth and allowing the shavings to pass through. As soon as the working pressure is released, the blade springs back to its normal position.

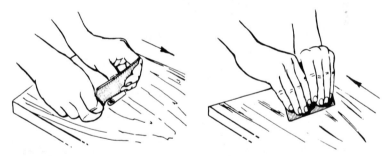

Fig. 24. A hand scraper and its method of use. In operation, the hand scraper is pushed or pulled, thus removing surface irregularities in the work; when properly used, it provides the work with a smooth and glossy finish.

jack plane is for very heavy, rough work. It can be used as a smoothing or jointer plane, which is a finer cut, if used properly.

Block planes are used primarily as a one-hand plane and almost exclusively for planing across the grain. The angle of the plane iron is different from other planes, and this is why it is generally called a low-angle block plane.

REVIEW QUESTIONS

1. What is a spokeshave?
2. Name the various types of planes.
3. What is a scraper?
4. What is a plane iron?
5. What is a low-angle block plane?

CHAPTER 16

Boring Tools

There are several kinds of boring tools; each class is adapted to perform different tasks, such as:

1. Punching.
2. Boring.
3. Drilling.
4. Countersinking.
5. Enlarging.

The various kinds of tools used for these operations are brad awls, gimlets and augers, drills, hollow augers and spur pointers, countersinks, and reamers. These tools are called bits when provided with a shank instead of a handle for use with a brace or for use in an electric drill press.

SCRATCH AWL

An awl is a pointed tool that is used for small starting holes for screws and nails. It can also be used to accurately scribe a line. Fig. 1 shows the scratch awl.

AUGERS

Augers are used for boring holes from ¼ inch to 2 inches. The sizes of auger and Forstner bits are listed in 16ths. Fig. 2 illustrates a comparison between auger, Forstner, and twist bits. When made with a shank for use in a brace, this style of auger is commonly called a bit, as shown in Fig. 3.

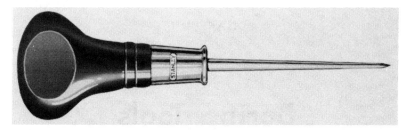

Fig. 1. A typical scratch awl.

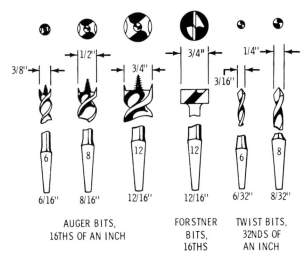

AUGER BITS,
16THS OF AN INCH

FORSTNER
BITS,
16THS

TWIST BITS,
32NDS OF
AN INCH

Fig. 2. Typical auger, Forstner, and twist bits and the methods of marking their size.

Because of the enormous variety of bits on the market, it is difficult to select the best one for a given purpose. Accuracy, speed, what? The bits look alike. For example, Fig. 4 shows an enlarged view of just two common styles of wood bits. One has a screw point, the other has a brad, or diamond, point. Note also that one style has a solid center with a single spiral running around it, while the other is a double-spiral twist bit. But they do look alike. The information given here will help, and so will experience.

It is not generally understood how important a part the screw

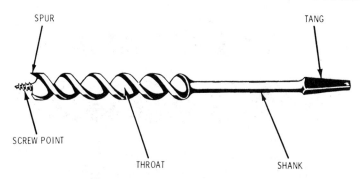

Fig. 3. A typical auger bit illustrating its component parts.

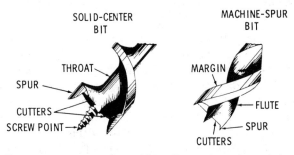

Fig. 4. Two common types of wood bits. The cutting edges of these bits are quite similar. The opening between the spiral is called the throat, but in some styles of double-twist bits, it is called the "flute." Both terms are used alternately and mean the same thing.

thread plays in boring. The terms "coarse" and "fine," as applied to a screw thread, are relative and may be applied to either single-thread or double-thread screws. The bit having a given number of double threads to the inch, provided the cutters are pitched to correspond with those threads, will bore just as fast as a bit with half that number of single threads to the inch, provided the cutters are of the same pitch. If the cutters have less pitch than the threads, they will act as a stop gauge, thereby keeping the bit from boring as fast as it would without such an obstruction.

Fig. 5 shows a hollow-spiral and two double-spur bits. The hollow-spiral bit has a screw point and only one cutting edge; its

191

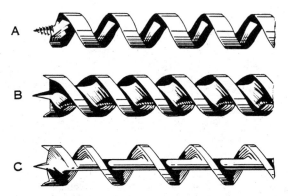

Fig. 5. Typical auger-bit styles. The auger bits shown represent the hollow-spiral bit (single twist) (A), the double-spur bit (fluted) (B), and the double-spur bit (solid center) (C), respectively.

hollow center permits the easy passage of wood chips, thereby allowing this type of bit to cut faster, especially when boring deep holes. The expansive bit, shown in Fig. 6, is so called because it may be set for various diameter holes within its capacity, thereby taking the place of many large bits.

Another excellent bit to use when boring large holes is the multispur type, shown in Fig. 7A; although this bit is relatively low in price, it does not readily bore holes of various diameters. A double-spur, twist-drill bit is shown in Fig. 7B; this is one of the cleanest and fastest cutting types of wood bits on the market. The flat-center bit, Fig. 7C, has only one cutting edge and is used for boring large shallow holes. Both the hole and the coun-

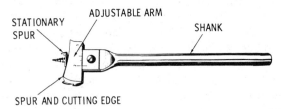

Fig. 6. A typical expansive auger bit with an adjusting screw. The expansive bit obsoletes the necessity for many large bits. The cutter may be adjusted for various size holes. The size of the hole to be cut may be reduced or enlarged ⅛ inch by turning the adjusting screw one complete revolution in the direction desired. Test for the correct size setting on a piece of waste wood before boring the hole in the wood being worked.

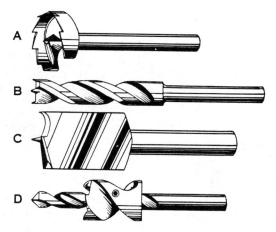

Fig. 7. Special wood-bit styles. The bits represented are a multispur bit (A), a double-spur bit (B), a center bit (C), and a countersink bit (D). Bits not equipped with feed-screw points are usually meant to be used in the drill press, whereas bits equipped with screw points are for use with the hand brace.

tersink can be cut in one operation with the countersink bit illustrated in Fig. 7D.

If it is necessary to drill holes to an exact depth, an adjustable bit gauge, such as the one shown in Fig. 8, may be used. This tool is simply a clamp that can be securely attached to any standard-size wood bit by means of two wing nuts, as illustrated.

Among other types of bits frequently used in woodworking shops are router bits, end cutters (for cutting rosettes, rounding, and shaping), and a variety of other designs. Bits may also be used to cut other soft materials, such as Plexiglas (Fig. 9).

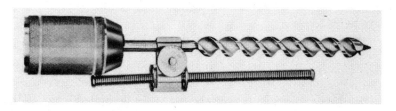

Fig. 8. A typical depth gauge and its use. An adjustable bit gauge of this type may be used to regulate the depth of the holes to be bored.

KVCC KALAMAZOO VALLEY COMMUNITY COLLEGE LIBRARY

Fig. 9. With the right bits hand drill can be used to make holes through a variety of materials such as Plexiglas.

It should be understood that the double-thread bit is intended for softwood, and the single-thread bit is intended for hardwood. The single-thread bit will not clog up as readily as the double-thread type; if the double-thread bit were left coarse enough so as not to clog, it would make the bit turn too hard.

To sharpen the spur of an auger, hold the bit in the left hand with the twist resting on the edge of the bench. Turn the bit around until the spur to be sharpened faces up. File the side of the spur next to the screw, carefully keeping the original bevel. File lightly until a fine burr shows on the outside, which is carefully removed by a slight stroke with a file; the result is a fine cutting edge.

To sharpen the cutter, hold the bit firmly in the left hand with the worn point down on the edge of the bench, slanted away from the hand with which you file. File from the inside back, and be careful to preserve the original bevel; take off the burr or rough edge. Never sharpen the outside of the spur.

It is rarely necessary or advisable to sharpen the worm; however, it may often be improved if it is battered by using a three-cornered file that is carefully manipulated; use a size that fits the thread. A half-round file is best for the lip and, with careful handling, may be used for the spur. Special auger-bit files are available for this purpose.

TWIST DRILLS

In addition to augers and gimlets, a carpenter should have a set of twist drills. These tools are used for drilling small holes where the ordinary auger or gimlet would probably split the wood. They come either with square shanks for use with bit braces or with straight shanks for use with electric drills, as illustrated in Fig. 10. These drills are available in standard sizes from $^1/_{16}$ to $^5/_8$ inch or more, varying in size by 32nds of an inch.

A twist drill differs from an auger or gimlet in that it has no screw and has a less acute cutting angle of the lip; therefore, there is no tendency to split the wood, since the tool does not pull itself in by a taper screw but enters by external pressure.

For many operations, especially where the smaller drills are used, as in drilling nail holes through boat ribs and planking, a geared breast drill is preferable to a brace.

COUNTERSINKS

Sometimes it is necessary to make a conical enlargement of a hole at the surface of the wood. This operation is performed by a bit tool called a countersink, which may be used in a hand brace,

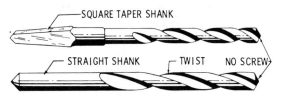

Fig. 10. A bit stock twist drill for use with a brace and a straight-shank twist drill for use in the drill press.

an electric drill, or a drill press. A typical countersink is shown in Fig. 11.

HAND DRILLS AND BRACES

The hand drill and brace, Fig. 12, are the conventional hand tools used by the carpenter for holding and turning bits. The hand drill is used for rapid drilling of small holes. The brace differs from the hand drill mainly in that with the brace the turning movement is applied directly to the bit by means of the handle swing, whereas the hand drill is equipped with a gear-pinion arrangement for turning the drill. The component parts of the hand drill and the brace are identified in Figs. 13 and 14, respectively.

Fig. 11. A typical rose countersink.

Fig. 12. A hand drill and a hand brace, respectively.

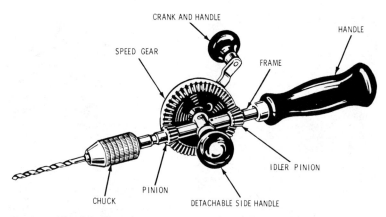

Fig. 13. The component parts of a typical hand drill.

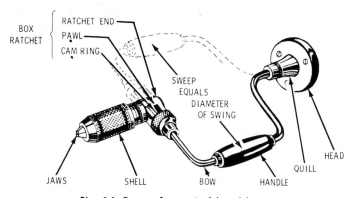

Fig. 14. Parts of a typical hand brace.

Satisfactory results in the use of boring tools are only obtained with practice and the use of good tools, each suitable for the particular job assigned to it. The work should be properly laid out, and the hole should be clearly marked. To bore a vertical hole, hold the brace and bit perpendicular to the surface of the work, as shown in Fig. 15. Compare the direction of the bit to the closest straightedge or to the sides of the vise. A try square may also be held near the bit to be certain of the true vertical position.

To bore a horizontal hole, hold the head of the brace cupped in the left hand against the stomach, with the thumb and

197

Fig. 15. The proper method of boring a vertical hole with the hand brace. The bit must be perpendicular to the work surface. Courtesy of Stanley.

forefinger around the quill, as shown in Fig. 16. To bore through the wood without splintering the second side, stop when the screw point reaches the other side, and finish the hole from that side. When boring with an expansive bit, it is best to clamp a piece of scrap wood to the second side and bore straight through.

Frequently, restricted working quarters make it necessary to use the ratchet device of the hand brace. The ratchet brace is indispensable when boring a hole in a corner or when a projecting object prevents the user from making a full turn with the handle. To actuate the ratchet, turn the cam ring. Turning the cam ring to the right will allow the bit to turn clockwise and give a ratchet action when the handle is turned left; turning the cam ring to the left will reverse this action.

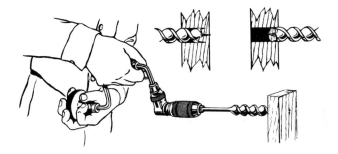

Fig. 16. The correct method of boring holes horizontally to prevent splitting and splintering.

SUMMARY

Various kinds of boring tools are used in woodworking shops. Tools like punches, drills, countersinks, enlargers, and boring implements are used every day in various operations. These tools are generally called bits when provided with a shank instead of a handle for use with a brace or for use in an electric drill.

There are various types and sizes of augers. Some augers have a screw point, and others have a brad, or diamond point. Fig. 5 illustrates the three different auger styles that are the most popular.

Twist drills are used for drilling small holes. They are designed with square shanks for brace or straight shanks for electric drills. The twist drill differs from an auger in that it has no screw and has a less acute cutting angle of the lip; therefore there is less danger of splitting the wood.

REVIEW QUESTIONS

1. What is a gimlet?
2. Explain the auger and its uses.
3. What is the difference between the flute and throat on a wood bit or auger?
4. What is the spur of a wood bit?
5. What is a reamer and how is it used?

CHAPTER 17

Fastening Tools

Classically, fastening tools refer to hammers, wrenches, screwdrivers, and the like that are used to install fasteners — screws and nails and the like that hold things together. But in relatively recent years there have been other tools, such as the rivet gun, staple gun, and Vise Grips, which more than deserve inclusion in any discussion of fastening tools. A roundup follows.

HAMMERS

The hammer is the all-important tool in carpentry, and there are numerous types to meet the varied conditions of use. All hammers worthy of the name are made of the best steel available and are carefully forged, hardened, and tempered.

The shapes of the claws of hammers vary slightly in the products of different manufacturers, though they may all be called curved-claw hammers. Carpenters often develop a preference. The straight-claw, or ripping, hammer is not as popular as the curved-claw hammer, shown in Fig. 1, with most workmen, because it does not grip nails for withdrawal quite so readily, and they cannot be withdrawn easily. The shape of the poll is immaterial, but the octagon-shaped poll seems to be most popular. All good hammers have slightly rounded faces, thereby making it possible to drive a nail head down flush with the wood without unduly marking the material.

Several types of hammers are currently available, as shown in Figs. 2 to 5. One popular type is forged (head and handle) of a single piece of drop-forged steel with the grip built up of leather

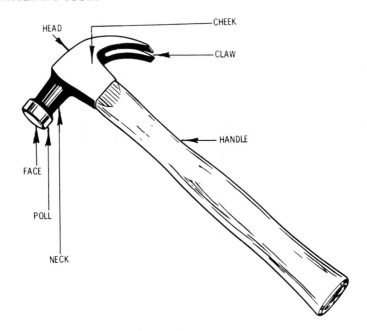

Fig. 1. A typical bell-faced nail hammer.

Courtesy Vaughan and Bushnell Mfg. Co.

Fig. 2. Full polished octagon neck, round face, air-cushioned neoprene grip hammer. This hammer is forged steel from head to toe with a hickory plug in the head to absorb shock.

or neoprene. Fiberglass-handle hammers have become popular also. The handle is made of polyester resin reinforced by continuous fiberglass in parallel form and has excellent strength. They are available in 13-, 16-, and 20-oz. nail types, as well as 16- and 20-oz. ripping types, with nonslip neoprene grips.

While metal and fiberglass hammers are popular, many workmen still prefer the old standard adze-eye hammer with a handle of springy hickory. True, such handles can become sprung, or

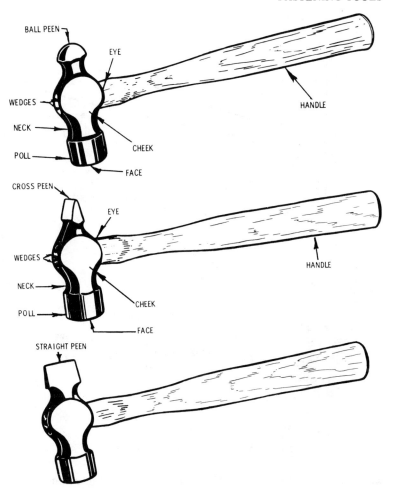

Fig. 3. Ball-, cross-, and straight-peen hammers.

perhaps break, but the "feel" of such hammers is something some craftsmen do not want to part with. Unquestionably though, anyone can become accustomed to the metal-handled tools in time.

It is not a good policy for the neophyte to change to a lighter hammer when doing trim work. He will usually be slow enough to acquire good control of one hammer. Builders and other

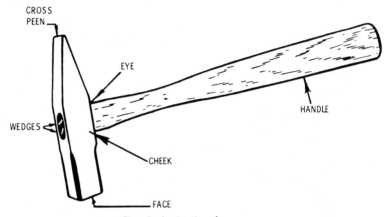

Fig. 4. A riveting hammer.

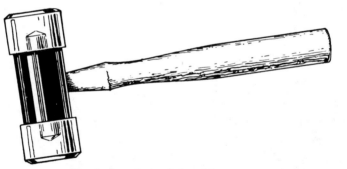

Fig. 5. A typical soft-faced hammer.

workmen who do a large amount of heavy spiking often use a 2-pound hammer, sometimes with a cut face. However, such hammers are of no use for anything else. The beginner should select a good hammer from an established and recognized manufacturer. The advice of an experienced foreman or carpenter is also extremely desirable. At any rate, the novice will bend a sufficient number of nails with the best of hammers until he acquires skill in using this most indispensable tool.

When using a hammer, the handle should be grasped a short distance from the end, and a few sharp blows rather than many light ones should be used when driving nails. Keep the hand and wrist level with the nailhead so that the hammer will hit the nail squarely on the head instead of at an angle. Failure to do this is the reason for difficulty so often experienced in driving nails straight.

The hammer face must be clean and dry to drive a nail straight; therefore, rub the face of the hammer frequently on wood. Hammers are designed to drive nails and not to hit wood (or fingers); thus, when starting, tap gently while the nail is guided with the fingers and finish with a nailset. Hammers vary in size from 5 to 20 ounces. Most carpenters and handymen use a hammer weighing 14 to 16 ounces.

It is always a good idea to protect eyes and face from dust and flying particles when using a hammer. Many safety goggles are designed to be worn over personal glasses, with large clear-vision lenses and air holes for maximum ventilation, as shown in Fig. 7.

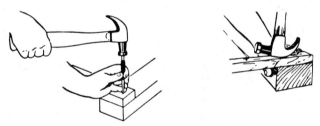

Fig. 6. When using hammer, hold it near the end. When driving nails, snap wrist, but hit nail squarely.

SCREWDRIVERS

A screwdriver is similar to a chisel and usually differs mainly in the working end, which is blunt. Few screwdrivers have an ideally shaped end. Usually the sides that enter the slot in the screw are tapered. This is done so that the end of the screwdriver will fit into screw slots of widely varying sizes, as shown in Fig. 8.

When using a screwdriver with a tapered blade tip, a force is set up due to the taper, which tends to push the end of the tool out of the slot. Therefore, it is better to have several sizes with

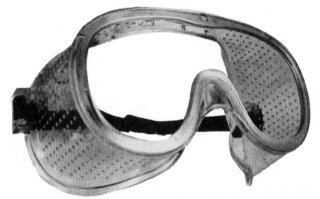

Courtesy Vaughan and Bushnell Mfg. Co.

Fig. 7. Safety goggles for use with striking tools.

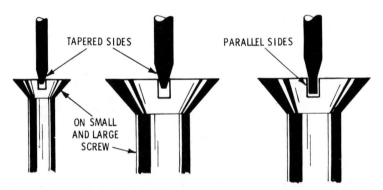

Fig. 8. The end of a screwdriver should be shaped so that its sides are parallel. A screwdriver whose end is tapered can be used, but considerable downward pressure must be exerted to prevent the screwdriver from rising out of the screw slot. With parallel sides, there is no tendency for the screwdriver to rise, no matter how much turning force is exerted.

properly shaped parallel sides than to depend on one size with tapered sides for all sizes of screws.

There are two general classes of screwdrivers—the plain and the so-called automatic. Fig. 9 shows a typical plain screwdriver. The operation of driving a screw with a plain screwdriver consists of giving it a series of half turns.

Where a number of screws are to be tightened, time can be saved by using a screwdriver bit which is used with a brace in

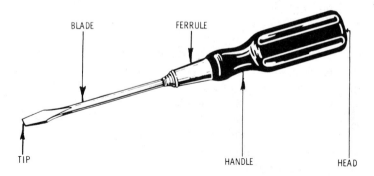

BLADE

FERRULE

TIP

HANDLE

HEAD

Fig. 9. A typical plain screwdriver.

the same manner as an auger bit. The quickest method of driving a screw is by means of the so-called automatic ratchet-type screwdriver, shown in Fig. 10. The advantage of this type over the plain screwdriver is that instead of grasping and releasing the handle from 25 to 30 times in turning a screw home, it is grasped once and with two or three back-and-forth strokes, the screw is driven home, thus saving labor and time. The screwdriver drives or withdraws screws according to the position of the ratchet shifter, because pressure on the handle causes the spindle and tip to rotate. The ratchet shifter can also lock the screwdriver into a rigid unit, so it can be used like a conventional screwdriver.

Special ratchet-type screwdrivers may be obtained with spi-

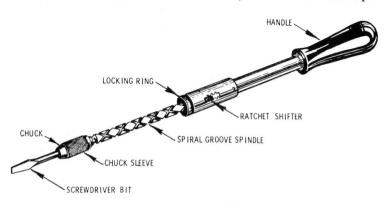

HANDLE

LOCKING RING

RATCHET SHIFTER

CHUCK

SPIRAL GROOVE SPINDLE

CHUCK SLEEVE

SCREWDRIVER BIT

Fig. 10. A spiral-ratchet screwdriver.

rals of different angles to suit working conditions, such as: a 40°
spiral for rapidly driving small screws, a 30° spiral for general
work, and a 20° spiral for driving large screws in hardwood.

The Phillips screwdriver, Fig. 11, is constructed with a specially shaped blade tip to fit Phillips cross-slot screws; the heads
of these screws have two perpendicular slots that intersect each
other in the center. This design checks the tendency of some
screwdrivers to slide out of the slot and on to the finished surface of the work. The Phillips screwdriver will not slip and burr
the end of the screw if the proper size tool is used.

The offset screwdriver, also shown in Fig. 11, is a handy tool
to use in tight corners where working room is limited. It usually
has one blade forged in line with the shank, or handle, but this is
not always the case; the other blade is at a right angle to the
shank. The ends of the screwdriver can be changed with each
turn, thereby working the screw into or out of its hole. This type
of screwdriver is normally used only when the screw location is
such that it prohibits the use of a plain or spiral-ratchet screwdriver. The offset screwdriver is also available in the ratchet form.

WRENCHES

The wrench, though it may not be so considered, can be a
somewhat dangerous tool, especially when great force is applied

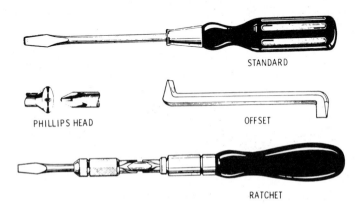

STANDARD

PHILLIPS HEAD   OFFSET

RATCHET

Fig. 11. A typical assortment of screwdrivers.

to move an obstinate nut. Under such conditions, the jaws often slip off the nut, thereby resulting in injury to the workman by violent contact with some metal part. You should be aware, then, of the potential problems.

There are three general classes of wrenches:

1. Plain.
2. Adjustable.
3. Socket.

Plain, or open-end, wrenches are of the solid nonadjustable type with fixed openings at each end, as shown in Fig. 12. A conventional set of wrenches for the carpenter's use contains between eight and twelve wrenches ranging in size from $5/16$ to 1 inch in jaw width.

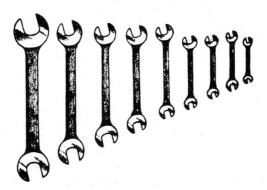

Fig. 12. Open-end wrenches.

Another form of plain wrench is the box-end wrench; these are so called because they "box," or completely surround, the nut or bolt head. Their usefulness is based to some extent on their ability to operate in close quarters. Because of their unique design, there is little chance of the wrench slipping off the nut. As shown in Fig. 13, twelve notches are arranged in a circle around the inside of the "box." This twelve-point wrench can be used to continuously loosen or tighten a nut with a handle rotation of only 15°, as compared to the 60° swing required by the open-end wrench if it is reversed after each swing. A combina-

Fig. 13. A typical box-end wrench.

tion wrench employs a box-end wrench at one end and an open-end wrench at the other; both ends are usually the same size, although this is not always the case.

Adjustable wrenches are somewhat similar to open-end wrenches; the primary difference is that one jaw is adjustable, as shown in Fig. 14. The angle of the opening with respect to the handle of an adjustable wrench is 22.5°. A spiral worm adjustment in the handle permits the width of the jaws to be varied from zero to ½ inch or more, depending on the size of the wrench. Always place the wrench on the nut so that the pulling force is applied against the stationary jaw. Tighten the adjusting screw so that the jaws fit the nut snugly.

The wrench most commonly found in a carpenter's toolbox is probably the old-fashioned monkey wrench, shown in Fig. 15. For most of his uses, this wrench fills the bill well. Fit the jaws of the wrench snugly on the nut to be turned; it is usually advis-

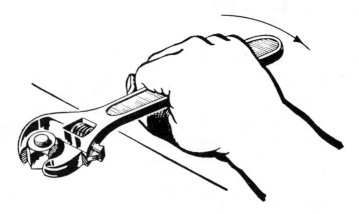

Fig. 14. The adjustable wrench and the method of tightening a nut.

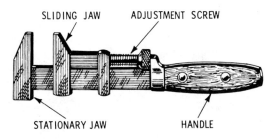

SLIDING JAW   ADJUSTMENT SCREW

STATIONARY JAW   HANDLE

Fig. 15. A typical monkey wrench.

able to turn *toward* the screw side of the handle, since the wrench is then not so likely to slip off the nut. Do not try to turn the nut or bolt with the tips of the jaws, or slip a piece of pipe over the handle to increase leverage; a stillson pipe wrench may be able to take it, but an ordinary monkey wrench won't. Keep the adjustment screw well oiled.

OTHER FASTENING TOOLS

As mentioned earlier, there are a number of other fastening tools to know about. The so-called "Pop" rivet gun is good for securing things where only one side is accessible. You just put a rivet in the head, stick the rivet in a hole that passes through both parts, and squeeze—the rivet clamps them together.

The hand staple gun (an electric model is also available) is great for rapid application of staples when fastening ceiling tile and the like; there is another stapler that is operated by being rapped against a surface, like a hammer, to set a staple. This is popular with insulation installers.

A close cousin of the stapler is the nailing gun, a recent addition to the hand-tool field that looks like a stapler and enables one to drive and countersink brads at the same time for easier installation of paneling.

Finally—and to some minds most important—there is the clamping pliers, or Vise Grips (a brand name). This great tool is wrench and clamp in one, and is excellent when you need to get purchase on a bolt or nut; it clamps on solid. It has many other uses, which will be discovered as the tool is used.

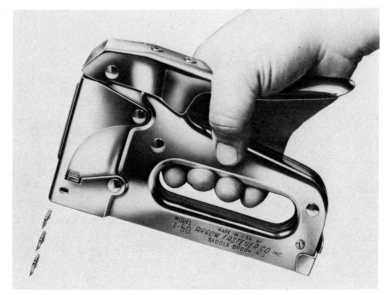

Fig. 16. A staple gun. Courtesy of Arrow Fastener Co.

SUMMARY

Fastening tools include various hammers, screwdrivers, and wrenches. These tools are used for securing or joining various parts of wood or other materials together with nails, screws, or bolts.

The hammer is a very simple striking tool and is made in numerous sizes and shapes to perform various tasks. All hammers worthy of the name are made of the best steel, carefully forged, hardened, and tempered. There are various groups or classifications, such as nail or claw, ball peen, soft face, cross and straight peen, and tinners and riveting. The 16-oz. size is most popular.

Screwdrivers are designed to insert or remove screws from various materials. There are several classes of screwdrivers used in the average shop, such as plain, Phillips, offset, and ratchet. The plain screwdriver is very similar to a chisel except it has a blunt end. The Phillips screwdriver is made with a specially shaped blade tip that fits Phillips cross-slot screws.

212

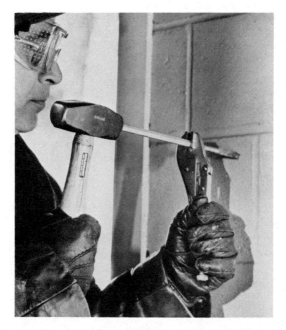

Fig. 17. Vise Grip pliers. Courtesy of Vaughn & Bushnell.

The wrench is a tool for tightening or loosening bolts and nuts used in the assembly of numerous articles of wood or other material. The majority of nuts and bolt heads are hexagonal (six-sided), although other shapes are sometimes encountered. The wrench is designed to grip these nuts and bolt heads and turn them by means of lever action exerted at the handle. Various types manufactured are open end, monkey, adjustable, box end, and combination. Other fastening tools include the rivet gun, staple gun, and Vise Grips (pliers).

REVIEW QUESTIONS

1. What is a Phillips screwdriver?
2. How are hammers sized for a particular job?
3. Name the various types of wrenches manufactured.
4. Name the various types of hammers manufactured.
5. What are the requirements for a good screwdriver?

213

CHAPTER 18

Sharpening Tools

As has been mentioned at a number of points in this book, keeping tools sharp is important not only for proper performance but for safety. You want to be able to know where the tool you're guiding is going to go.

For the sharpening job, many items are available, as follows.

GRINDING WHEELS

The composition of a grinding wheel consists of the cutting material or abrasive (usually called "grit") and the bond. The cutting quality of a wheel depends chiefly on the grit and the hardness of the bonding material. The object of the bond is not only to hold the particles of grit together with the proper degree of safety but also to vary its tensile strength. The grinding wheel is called *hard* or *soft* depending on the tenacity with which the bond holds the particles together. A grinding wheel is said to be too hard when the bond retains the surface or cutting particles until they become dull, and it is said to be too soft when the particles are not held long enough to prevent undue wear of the wheel. Wheels are bonded by the vitrified, silicate, elastic, and rubber processes. The size of the grinder, such as the one shown in Fig. 1, is commonly taken from the diameter of the abrasive wheel or wheels. For example, a grinder with a 7-inch-diameter wheel is called a "seven-inch grinder."

Abrasive wheels are normally graded with respect to their "grit capacity." A grinding wheel that is to be used for sharpening tools should preferably be an aluminum oxide wheel of ap-

Fig. 1. Five-inch bench grinder. Courtesy of Black & Decker.

proximately grade 60 grit and should be of a minimum hardness. A coarse, soft-grit wheel can remove material more rapidly than one with a finer grit, but the surface produced on the edge of the tool will be rougher than that produced with the fine-grit wheel. Therefore, to be suitable for grinding woodworking tools, a grinding wheel should be soft and should have a fine grit.

When using abrasive wheels for grinding woodworking tools, the high surface speed of the wheel in contact with the tool generates considerable heat; therefore, to reduce this heat production, the tool should be held lightly against the wheel and should frequently be dipped in water. If these two precautions are not taken, the edge of the tool may be severely burned, thereby reducing or even eliminating its usefulness. If a wheel that is running at a nominal speed has a tendency to burn the tool being sharpened, the wheel should be closely examined to determine whether it needs dressing.

OILSTONES

Oilstones are used after the grinding operation to give the tool the highly keen edge necessary to cut wood smoothly. The oilstone is so called because oil is used on a whetting stone to carry off the heat resulting from friction between the stone and the tool and to wash away the particles of stone and steel that are worn off in the sharpening process. The process of rubbing the tool on the stone is called "honing."

There are a number of other useful devices that can make the sharpening job simpler. For example, to sharpen and hone a plane there is a honing guide that allows you to clamp the plane iron over the honing stone at the angle you wish. Rubbing the iron back and forth sharpens it.

To sharpen bits, there are jigs that allow you to hold the bit at the correct angle on the grinding wheel.

Additionally, one manufacturer, Black and Decker, sells a drill bit sharpener that works like a pencil sharpener. You stick the bit into the opening in the device and turn it on. The bit is sharpened to the proper angle automatically.

SUMMARY

The grinder consists essentially of a horizontal spindle, the ends of which are threaded and fitted with flanges to take the grinding wheel. The spindle is either direct driven or has a conventional belt drive. The composition of a grinding wheel consists of the cutting material or abrasive, and the bond. The cutting quality of the grinding wheel depends on the grit and the hardness of the bonding material.

Oilstones are used after the grinding operation to give the tool the very keen edge necessary to cut wood smoothly. The oilstone is so called because oil is used to carry off the heat resulting from friction between the stone and the tool. There are natural stones and artificial stones, grouped according to the locality in which they are found.

A variety of other devices are available for sharpening individual items.

REVIEW QUESTIONS

1. What are the two operations involved in sharpening wood-working tools?
2. Why are oilstones used in tool sharpening?

CHAPTER 19

How To Sharpen Tools

Sharpening tools, you should know, is not as easy as it may appear—at least when it is done correctly. If you want to, you can have your tools sharpened professionally for a very small fee. On the other hand, most craftsmen want to learn how to sharpen tools for two reasons: The tools can be sharpened at the craftsman's convenience, and it is a very satisfying experience.

GRINDING

First, the tool is placed on a grinding wheel in order to bring the bevel to the correct angle and to grind out any nicks that may exist in the cutting edge. Although this takes out the nicks and irregularities that are visible to the eye, the edge is still rough, as seen under a microscope. This roughness is considerably reduced by honing on an oilstone, although it is impossible to make the edge perfectly smooth because of the granular structure of the material.

When grinding tools on a stone without the use of either a guide or a rest, the tool is firmly pressed to and held at an angle of approximately 60° on the face of the revolving stone with both hands. Do not apply too much pressure, especially when grinding with the rapidly revolving emery wheels, since it is all too easy to burn the temper out of the tool. The edge of the tool must be continuously watched, especially with dry wheels. In case of a dry wheel, the tool must be immersed frequently in water to prevent overheating.

Plane-iron cutters vary in their make, temper, quality of steel, and uses, and they must be ground and sharpened for the sort of work they are intended to execute. As previously explained, it is usual to grind a jack plane iron slightly curved, and a jointer or smoothing plane iron flat, except at the corners (see Fig. 13 of Chapter 15).

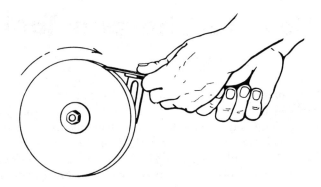

Fig. 1. Grinding is the first step in the sharpening process. Different tools require different techniques. For a plane iron, the grinding wheel should turn toward iron.

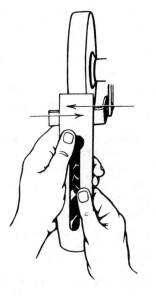

Fig. 2. Moving plane from side to side grinds bevel down completely.

Before condemning any plane iron, therefore, carefully measure and compare the bevel of the cut and the thickness of the cutter. If the bevel is too long, the plane will jump and chatter. If it is too short, it will not cut, so it must be ground to a proper bevel. The length of the bevel should be twice the thickness of the iron (see Fig. 18 of Chapter 15).

Hatchets, axes, and adzes are always ground to their proper bevels; some have double and others have single bevels (see Fig. 4 of Chapter 14). When grinding, the blade is held to the stone surface with the right hand, and the handle is held with the left hand and on the left side, reversing the tool as the opposite side is being ground or sharpened.

Drawknives and spokeshave cutters should be held with both hands, and the blade should be kept horizontal on the stone as it is revolved toward the operator. Some woodworkers prefer to grind with the stone rotating toward the cutting edge, while others prefer to grind with the stone rotating away from the cutting edge. The latter is the safer method, because with the stone advancing, there is a danger of injury to the operator in case the tool digs into the stone. Do not use too much pressure with an advancing stone.

After the tool is ground, it should be honed.

HONING

After a tool has been ground on a grinding wheel, it will still have a wire edge. This edge must be removed, and the edge must be further treated by honing on an oilstone (see Fig. 19 of Chapter 15). The oilstone is constantly needed during all operations in carpentry in which the plane and chisel are used. It is needed more frequently than the grinding wheel, because the grinding wheel is only necessary when the tool becomes nicked, or when the edge becomes too dull to be sharpened on the oilstone without an undue amount of labor. The size of a typical oilstone for general use is approximately $2'' \times 8''$ or $2'' \times 9''$.

One rather desirable stone is the double carborundum, that is, a carborundum oilstone with one side coarse and the other side fine. Begin to hone on the coarse side and finish on the fine side with this type of stone. It is absolutely necessary to keep the

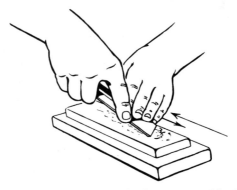

Fig. 3. After grinding, oilstone is used. Plane is moved back and forth on stone with back edge slightly raised. Stone is kept moist with oil.

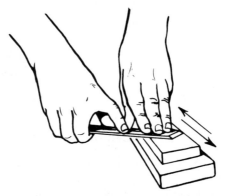

Fig. 4. Wire or feather edge is removed with plane iron held flat on stone.

oilstone clean and in perfect condition. The simple reality is that a stone that isn't in good condition won't work properly. Oilstones should always be kept in their case when not in use. Use only a thin, clear oil on oilstones, and wipe the stone clean after using. Then moisten the stone with clean oil.

To clean an oilstone, wash it in kerosene; this will remove the gummed surface oil. This may be more easily and thoroughly done by heating the stone on a hot plate. A natural stone may also be heated on a hot plate to remove the surplus or gummed

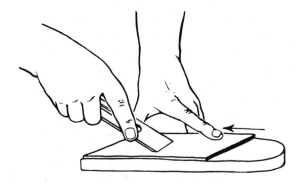

Fig. 5. Move stone back and forth, as shown, to finish.

oil, after which a good cleaning with gasoline or ammonia will usually restore its cutting qualities; if this treatment does not work, scour the stone with a piece of loose emery or sandpaper which has been fastened to a perfectly smooth board. Clean the stone in a well ventilated area and observe safety precautions.

When applying chisels and plane irons to an oilstone, the tool is held face up with both hands, the left hand in front, palm up, with the thumb on top, the fingers grasping the tool from underneath. The right hand is held behind the left hand, palm down, with the thumb under and the fingers reaching across the face of the tool. The blade edge is then moved back and forth with a sliding rotary motion on the face of the stone (which is first lubricated with oil or water). The blade angle is generally ap-

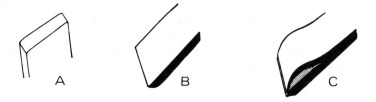

Fig. 6. A: When sharpening a plane iron, be aware that plane marks will show less on a surface if the iron's corners are slightly rounded. B: When sharpening, do not rock the iron. A round bevel will occur that doesn't cut well. C: A bevel on the flat side of the iron prevents the cap iron from fitting tightly, and shavings can clog the plane.

proximately 60°. After ten or twelve "rubs," the blade is turned over and rubbed flat on the face side. The blade is then stropped; this may be done either by a slapping action or rubbing on a piece of old belting or leather set on top of the oilstone case. When this is done, the keenness of the blade may be tested with the thumb or by drawing the edge across the thumbnail, but this test must be done carefully to avoid injury.

Outside gouges are sharpened in the same manner as chisels. The tool should be rolled forward and backward when grinding the bevel. A whetstone is used to remove the wire edge by rubbing on the inside concave surface; the curved edge of the whetstone must fit exactly the arc of each gouge as closely as possible. Inside gouges must be ground on a curved stone and whetted to keen edges with the oilstones and whetstones.

Hollows and rounds, beading, and other special plane cutters are usually sharpened with whetstones and rarely require grinding. If they become nicked or injured on their edges, they are utterly useless.

Cold chisels, punches, and nailsets are best sharpened or pointed on grinding wheels. Carving tools are sharpened with small, fine whetstones.

When honing or whetting fine bench chisels, the burnished-face side must be kept perfectly flat on the face of the oilstone by pressing firmly down with the fingers of the left hand; the handle is held in the right hand. The rubbing action must be gentle and must also be rapidly repeated, turning the tool over constantly.

The edge of the chisel blade may slope slightly to the side of the oilstone, and it should be moved back and forth in a rotary motion on the stone. Do not raise the angle of the chisel too high on the stone, or the chisel will dig into and damage the surface of the oilstone. The oilstone should be wiped clean and reoiled frequently if several tools are to be sharpened.

SUMMARY

It is always important to keep a good working edge on woodworking tools in order to do satisfactory work. The cutting edge must always be free from nicks and have the proper bevel.

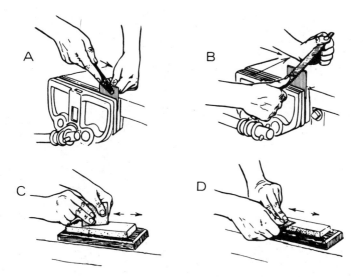

Fig. 7. A: To sharpen scraper blade on spokeshave or any bevel-edge blade, first remove old burr by running a mill file against flat edge on flat side. B: File or grind the bevel at about 45°. Push file forward and to the side. C: Whet the bevel side of blade on oilstone. D: Knock off burrs by rubbing face side against stone.

Sharpening is done by subjecting the tools to grinding and honing.

The tools are placed on a grinding wheel in order to bring the bevel to the correct angle and to grind out the nicks and irregularities. After the grinding process the tool is then honed on an oilstone to remove the roughness.

REVIEW QUESTIONS

1. What angle is required on most cutting blades?
2. Explain the process of grinding the tools.
3. How do you clean an oilstone?
4. What is honing?

CHAPTER 20

How To Use the Steel Square

On most construction work, especially in house framing, the "steel square" is invaluable for accurate measuring and for determining angles. The correct name is *framing square,* because the square with its markings was designed especially for marking timber in framing. Whatever the name, the square, when used properly, is a wonderful tool.

The tool, with its various scales and tables, has been explained in Chapter 7. The goal of this chapter is to explain these markings in more detail and also to explain their application by examples showing actual uses of the square. The following names are commonly used to identify the different portions of the square and should be noted and remembered:

Body—The long, wide member.
Face—The sides visible (both body and tongue) when the square is held by the tongue in the right hand with the body pointing to the left (see Fig. 1).
Tongue—The short, narrow member.
Back—The sides visible (both body and tongue) when the square is held by the tongue in the left hand with the body pointing to the right (see Fig. 1).

The square most generally employed has an 18-inch tongue and a 24-inch body. The body is 2 inches wide, and the tongue is

1½ inches wide, $^3/_{16}$ inch thick at the heel or corner for strength, diminishing, for lightness, to the two extremities to approximately $^3/_{32}$ inch. The various markings on squares are of two kinds:

1. Scales, or graduations.
2. Tables.

When buying a square, it is advisable to get one with all the markings rather than a "budget" unit on which the manufacturer has omitted some of the scales and tables. The following comparison illustrates the difference between an incomplete and a complete square.

The square with the complete markings will cost more, but in the purchase of tools, you should make it a rule to purchase only the finest made. The general arrangements of the markings on squares differ somewhat with different makes; it is advisable to examine the different makes before purchasing to select the one best suited to your specific requirements.

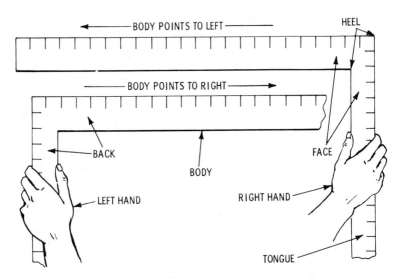

Fig. 1. The face and back sides of a framing square. The body of the square is sometimes called the blade.

APPLICATION OF THE SQUARE

As stated previously, the markings on squares of different makes sometimes vary both in their position on the square and the mode of application. However, a thorough understanding of the application of the markings on any first-class square will enable the student to easily acquire proficiency with any other square.

	Tables	Graduations
Cheap Square	Rafter, Essex, Brace	1/16, 1/12, 1/8, 1/4
Complete Markings	Rafter, Essex, Brace, Octagon, Polygon cuts	1/100, 1/64, 1/32, 1/16, 1/12, 1/10, 1/8, 1/4

Scale Problems

The term "scales" is used to denote the inch divisions of the tongue and body length found on the outer and inner edges; the inch graduations are divided into $\frac{1}{4}$, $\frac{1}{8}$, $\frac{1}{10}$, $\frac{1}{12}$, $\frac{1}{16}$, $\frac{1}{32}$, $\frac{1}{64}$, and $\frac{1}{100}$. All these graduations should be found on a first-class square. The various scales start from the heel of the square, that is, at the intersection of the two outer, or two inner, edges.

A square with only the scale markings is adequate to solve many problems that arise when laying out carpentry work. An idea of its range of usefulness is shown in the following problems.

Problem 1—To describe a semicircle given the diameter.

Drive brads at the ends of the diameter LF, as shown in Fig. 2. Place the outer edges of the square against the nails, and hold a lead pencil at the outer heel M; any semicircle can then be described, as indicated. This is the outer-heel method, but a better guide for the pencil is obtained by using the inner-heel method, which is also shown in the figure.

Problem 2—To find the center of a circle.

Lay the square on the circle so that its outer heel lies in the circumference. Mark the intersections of the body and tongue with the circumference. The line that connects these two points is a diameter. Draw another diameter (obtained in the same way); the intersection of the two diameters is the center of the circle, as shown in Fig. 3.

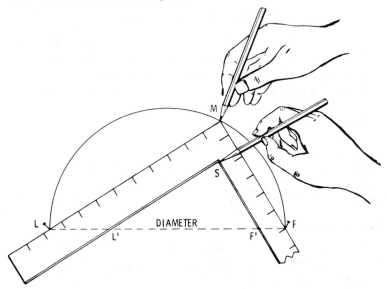

Fig. 2. Problem 1. *The outer-heel method is described in the text. For the inner-heel method, the pencil is held at S, and the distance L'F' should be taken to equal the diameter, with the inner edges of the square sliding on the brads.*

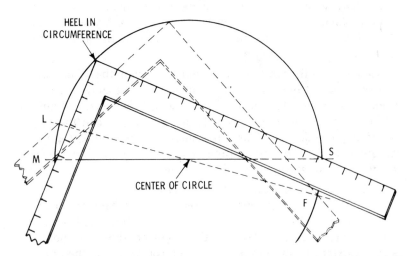

Fig. 3. Problem 2. *Draw diameters through points LF and MS, where the sides of the square touch the circle with the heel in the circumference. The intersection of these two lines is the center of the circle.*

Problem 3—To describe a circle through three points which are not in a straight line.

Join the three points with straight lines; bisect these lines, and, at the points of bisection, erect perpendiculars with the square. The intersection of these perpendiculars is the center from which a circle may be described through the three points, as shown in Fig. 4.

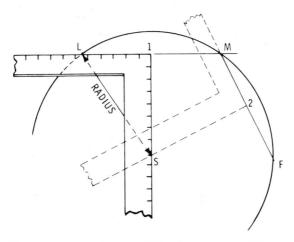

Fig. 4. Problem 3. Let points L, M, and F be three points which are not in a straight line. Draw lines LM and MF, and bisect them at points 1 and 2, respectively. Apply the square with the heel at points 1 and 2, as shown; the intersection of the perpendicular lines thus obtained, point S, is the center of the circle. Lines LS, MS, and FS represent the radius of the circle, which may now be described through points L, M, and F.

Problem 4—To find the diameter of a circle whose area is equal to the sum of the areas of two given circles.

Lay off on the tongue of the square the diameter of one of the given circles, and on the body the diameter of the other circle. The distance between these points (measure across with a 2-foot rule) will be the diameter of the required circle, as shown in Fig. 5.

Problem 5—To lay off angles of 30° and 60°.

Mark off 15 inches on a straight line, and lay the square so that the body touches one end of the line and the 7½-inch mark on the tongue is against the other end of the line, as shown in

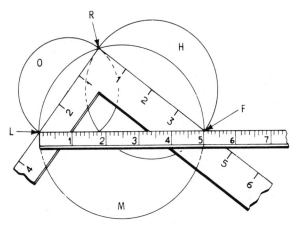

Fig. 5. Problem 4. *Let O and H be the two given circles, with their diameters LR and RF at right angles. Suppose the diameter of O is 3 inches and the diameter of H is 4 inches. Points L and F, at these distances from the heel of the square, will be 5 inches apart, as measured with a 2-foot rule. This distance LF, or 5 inches, is the diameter of the required circle. Proof:*
$$(LF)^2 = (LR)_2 + (RF)_5, \text{ or } 25 = 9 + 16.$$

Fig. 6. The tongue will then form an angle of 60° with the line, and the body will form an angle of 30° with the line.

Problem 6—To lay off an angle of 45°.

The diagonal line connecting equal measurements on either

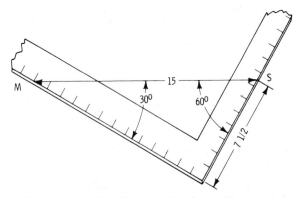

Fig. 6. Problem 5. *Draw line MS, 15 inches long. Place the square so that point S touches the tongue 7½ inches from the heel and point M touches the body. The triangle thus formed will have an angle of 30° at M and an angle of 60° at S.*

arm of the square forms angles of 45° with the blade and tongue, as shown in Fig. 7.

Problem 7—To lay off any angle.

Table 1 gives the values for measurements on the tongue and the body of the square so that by joining the points corresponding to the measurements, any angle may be laid out from 1° to 45°, as explained in Fig. 8.

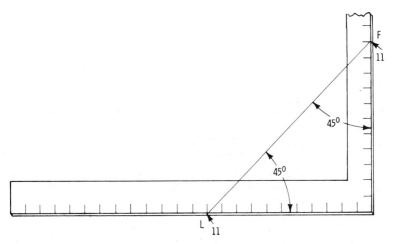

Fig. 7. Problem 6. Take equal measurements L and F on the body and tongue of the square. The triangle thus formed will have an angle of 45° at L and at F.

Problem 8—To find the octagon of any size timber.

Place the body of a 24-inch square diagonally across the timber so that both extremities (ends) of the body touch opposite edges. Make a mark at 7 inches and 17 inches, as shown in Fig. 9. Repeat this process at the other end, and draw lines through the pairs of marks. These lines show the portion of material that must be taken off the corners.

The side of an inscribed octagon can be obtained from the side of a given square by multiplying the side of the square by 5 and dividing the product by 12. The quotient will be the side of the octagon. This method is illustrated in Fig. 10.

The side of a hexagon is equal to the radius of the circumscribing circle. If the side of a desired hexagon is given, arcs should

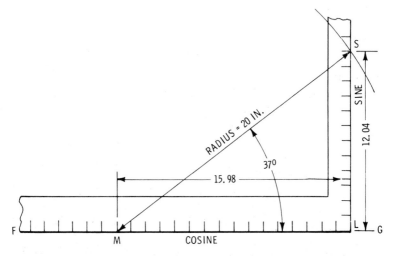

Fig. 8. Problem 7. Let 37° be the required angle. Place the body of the square on line FG, and, from Table 1, lay off LS (12.04) on the tongue and LM (15.98) on the body. Draw line MS; then angle LMS = 37°. Line MS will be found to be equal to 20 inches for any angle, because the values given in Table 1 for LS and MS are natural sines and cosines multiplied by 20.

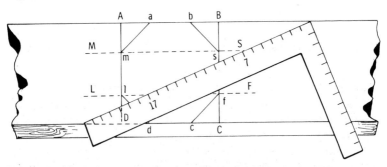

Fig. 9. Problem 8. Lay out square ABCD. Place the body of a 24-inch square as shown, and draw parallel lines MS and LF through points 7 and 17. These lines intercept sides ml and sf of the octagon. To lay off side sb, place the square so that the tongue touches point S and the body touches l, with the heel touching line AB. The remaining sides are obtained in a similar manner.

be struck from each extremity at a radius equal to its length. The point where these arcs intersect is the center of the circumscribing circle, and having described it, it is sufficient to lay off

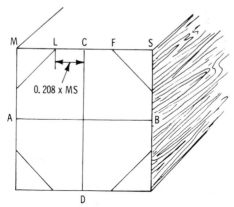

Fig. 10. Problem 8 (second method). Let lines AB and CD be center lines, and let line MS be one side of the square timber. Multiply the length of the side by 0.208; the product is half the side of the inscribed octagon. Therefore, lay off CF and CL, each 0.208 times side MS; LF is then one side of the octagon. Set dividers to distance CL, and lay off the other sides of the octagon from the center lines to complete the octagon.

cords on its circumference equal to the given side to complete the hexagon.

Square-and-Bevel Problems

By the application of a large bevel to the framing square, the combined tool becomes a calculating machine, and by its use, arithmetical processes are greatly simplified. The bevel is preferably made of steel blades. The following points should be observed in its construction:

The edges of each blade must be true; blade E in Fig. 11 must lie under the square so that it does not hide the graduations; the two blades must be fastened by a thumbscrew to lock them together; blade L should have a hole near each end and one in the middle, so that blade E may be shifted as required, with a large notch near each hole in order to observe the position of blade E.

Problem 9—To find the diagonal of a square.

Set blade E to 10⅜ on the tongue and to 15 on the body. Assume an 8-inch square. Slide the bevel sideways along the

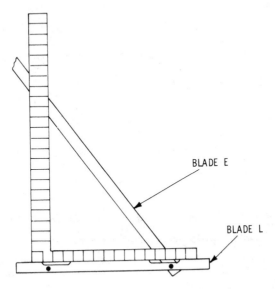

Fig. 11. The application of a bevel to the square for solving square-and-bevel problems.

Table 1. Angle Table for the Square

Angle	Tongue	Body	Angle	Tongue	Body	Angle	Tongue	Body
1	0.35	20.00	16	5.51	19.23	31	10.28	17.14
2	0.70	19.99	17	5.85	19.13	32	10.60	16.96
3	1.05	19.97	18	6.18	19.02	33	10.89	16.77
4	1.40	19.95	19	6.51	18.91	34	11.18	16.58
5	1.74	19.92	20	6.84	18.79	35	11.47	16.38
6	2.09	19.89	21	7.17	18.67	36	11.76	16.18
7	2.44	19.85	22	7.49	18.54	37	12.04	15.98
8	2.78	19.81	23	7.80	18.40	38	12.31	15.76
9	3.13	19.75	24	8.13	18.27	39	12.59	15.54
10	3.47	19.70	25	8.45	18.13	40	12.87	15.32
11	3.82	19.63	26	8.77	17.98	41	13.12	15.09
12	4.16	19.56	27	9.08	17.82	42	13.38	14.89
13	4.50	19.49	28	9.39	17.66	43	13.64	14.63
14	4.84	19.41	29	9.70	17.49	44	13.89	14.39
15	5.18	19.32	30	10.00	17.32	45	14.14	14.14

tongue until blade E is against point 8. The other edge will touch $11^5/_{16}$ on the body; this is the required diagonal.

Problem 10—To find the circumference of a circle from its diameter.

Set the bevel blade to 7 on the tongue of the square and to 22 on the body. The reading on the body will be the circumference corresponding to the diameter at which E is set on the tongue. To reverse the process, use the same bevel, and read the required diameter from the tongue, the circumference being set on the body.

Problem 11—Given the diameter of a circle, find the side of a square of equal area.

Set the bevel blade to 10⅝ on the tongue and to 12 on the body. The diameter of the circle, on the body, will give the side of the equal square on the tongue. If the circumference is given instead of the diameter, set the bevel to 5½ on the tongue and to 19½ on the body, thereby finding the side of the square on the tongue.

Problem 12—Given the side of a square, find the diameter of a circle of equal area.

Using the same bevel as in Problem 11, blade E is set to the given side on the tongue of the square, and the required diameter is read off the body.

Problem 13—Given the diameter of the pitch circle of a gear wheel and the number of teeth, find the pitch.

Take the number of teeth, or a proportional part, on the body of the square and the diameter, or a similar proportional part, on the tongue, and set the bevel blade to those marks. Slide the bevel to 3.14 on the body, and the number given on the tongue multiplied by the proportional divisor will be the required pitch.

Problem 14—Given the pitch of the teeth and the diameter of the pitch circle in a gear wheel, find the number of teeth.

Set the bevel blade to the pitch on the tongue and to 3.14 on the body of the square. Move the bevel until it marks the diameter on the tongue. The number of teeth can then be read from the blade. If the diameter is too large for the tongue, divide it and the pitch into proportional parts, and multiply the number found by the same figure.

Problem 15—The side of a polygon being given, find the radius of the circumscribing circle.

Set the bevel to the pairs of numbers in Table 2, taking one-eighth or one-tenth of an inch as a unit. The bevel, when locked, is slid to the given length of the side, and the required length of the radius is read on the other leg of the square. For example, if a pentagon (5 sides) must be layed out with a side of 6 inches, the bevel is set to the figures in column 5 with the lesser number set on the tongue. In this case, $^{74}/_8$ = 9¼ on the tongue, and 8⅞ = 10⅞ on the body of the square. Slide the bevel to 6 on the body. The length of the radius, $5^3/_{32}$, will be read on the tongue.

Problem 16—To divide the circumference of a circle into a given number of equal parts.

From the column marked Y in Table 3, take the number opposite the given number of parts. Multiply this number by the

Table 2. Inscribed Polygons

Number of Sides	3	4	5	6	7	8	9	10	11	12
Radius	56	70	74	60	60	98	22	89	80	85
Side	97	99	87	60	52	75	15	95	45	44

Table 3. Cords or Equal Parts

No. of Parts		Y	Z	No. of Parts	Y	Z	No. of Parts	Y	Z
3	Triangle	1.732	.5773	15	.4158	2.4050	40	.1569	6.3728
4	Square	1.414	.7071	16	.3902	2.5628	45	.1395	7.1678
5	Pentagon	1.175	.8006	17	.3675	2.7210	50	.1256	7.9618
6	Hexagon	1.000	1.0000	18	.3473	2.8793	54	.1163	8.5984
7	Heptagon	.8677	1.1520	19	.3292	3.0376	60	.1047	9.5530
8	Octagon	.7653	1.3065	20	.3129	3.1962	72	.0872	11.462
9	Nonagon	.6840	1.4619	22	.2846	3.5137	80	.0785	12.733
10	Decagon	.6180	1.6184	24	.2610	3.8307	90	.0698	14.327
11	Undecagon	.5634	1.7747	25	.2506	3.9904	100	.0628	15.923
12	Duodecagon	.5176	1.9319	27	.2322	4.3066	108	.0582	17.182
13	Tridecagon	.4782	2.0911	30	.2090	4.7834	120	.0523	19.101
14	Tetradecagon	.4451	2.2242	36	.1743	5.7368	150	.0419	23.866

238

radius of the circle. The product will be the length of the cord to lay off on the circumference.

Problem 17—Given the length of a cord, find the radius of the circle.

This is the same as Problem 16, but the present form may be more expeditious for calculations. The method is useful for determining the diameter of gear wheels when the pitch and number of teeth have been given. Multiply the length of the cord, width of the side, or pitch of the tooth by the figures found corresponding to the number of parts in column Z of Table 3. The result is the radius of the desired circle.

TABLE PROBLEMS

The term "table" is used here to denote the various markings on the framing square with the exception of the scales already described. Since these tables relate mostly to problems encountered in cutting lumber for roof-frame work, it is first necessary to know something about roof construction so as to be familiar with the names of the various rafters and other parts. Fig. 12 is a view of a roof frame showing the various members. In the figure it will be noted that there is a plate at the bottom and a ridge timber at the top; these are the main members to which the rafters are fastened.

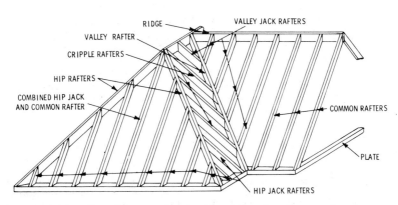

Fig. 12. A typical roof frame, showing the ridge, the plate, and various types of rafters.

Main or Common Rafters

The following definitions relating to rafters should be carefully noted:

The **rise** of a roof is the distance found by following a plumb line from a point on the central line of the top of the ridge to the level of the top of the plate.

The **run** of a common rafter is the shortest horizontal distance from a plumb line through the center of the ridge to the outer edge of the plate.

The **rise per foot run** is the basis on which rafter tables on some squares are made. The term is self-defining. Other roof components are illustrated in Fig. 13.

To obtain the rise per foot run, multiply the rise by 12 and divide by the run; thus:

$$rise \ per \ foot \ run = \frac{rise \times 12}{Run}$$

The factor 12 is used to obtain a value in inches, since the rise and run are normally given in feet.

Example—If the rise is 8 feet and the run is 8 feet, what is the rise per foot run?

$$rise \ per \ foot \ run = \frac{8 \times 12}{8} = 12 \ inches$$

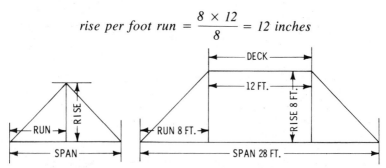

Fig. 13. The terms rise, run, span, and deck are illustrated in two types of roofs. If the rafters rise to a deck instead of a ridge, subtract the width of the deck from the span. For example, assume the span is 28 feet and the deck is 12 feet; the difference is 16 feet, and the pitch = 8/(28 − 12) = ½.

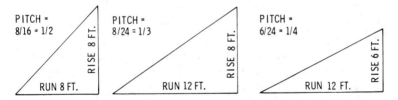

Fig. 14. To obtain the pitch of any roof divide the rise of the rafters by twice the run.

The rise per foot run is always the same for a given pitch and can be readily remembered for all ordinary pitches; thus:

Pitch ....½ ⅓ ¼ ¹/₆

Rise per foot run (in.) 12 8 6 4

The pitch can be obtained if the rise and run are known, as shown in Fig. 14, by dividing the rise by twice the run, or

$$pitch = \frac{rise}{2 \times run}$$

In roof construction, the rafter ends are cut with slants which rest against the ridge and the plate, as shown in Fig. 15A. The slanting cut which rests against the ridge board is called the *plumb*, or *top*, cut, and the cut which rests on the plate is called the *seat*, or *heel*, cut.

The length of the common rafter is the length of a line from the outer edge of the plate to the top corner of the ridge board or, if there is no ridge board, from the outer edge of the plate to the vertical center line of the building, as shown in Fig. 15B. The run of the rafter, then, in the first case is one-half the width of the building less one-half the thickness of the ridge, if any; if there is no ridge board, the run is one-half the width of the building. Where there is a deck, the run of the rafters is one-half the width of the building less one-half the width of the deck.

Now, with a 24-inch square, draw diagonals connecting 12 on the tongue (corresponding to the run) to the value from Table 4 on the body (corresponding to the rise) to obtain the pitch angle

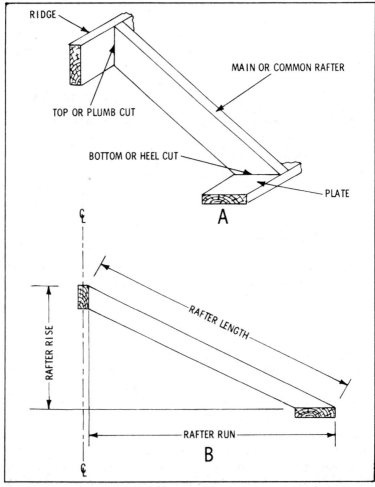

Fig. 15. A portion of the roof frame, showing the top, or plumb, cut and the bottom, or heel, cut is illustrated in A. The length of a common rafter is shown in B.

Table 4. Pitch Table

Pitch	1	11/12	5/6	3/4	2/3	7/12	1/2	5/12	1/3	1/4	1/6	1/12
Run	12	12	12	12	12	12	12	12	12	12	12	12
Rise	24	22	20	18	16	14	12	10	8	6	4	2

for any combination of run and rise. This procedure is further illustrated in Fig. 16.

Hip Rafters

The hip rafter represents the hypotenuse, or diagonal, of a right-angle triangle; one side is the common rafter, and the other side is the plate, or that part of the plate lying between the foot of the hip rafter and the foot of the adjoining common rafter, as shown in Fig. 17.

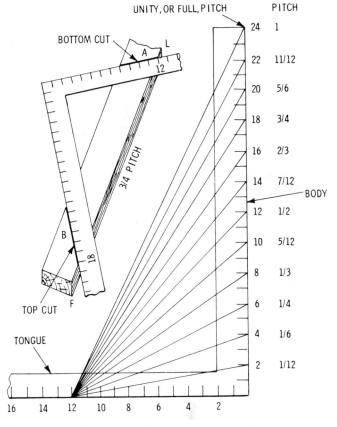

Fig. 16. The application of the framing square for obtaining the various pitches given in Table 4.

243

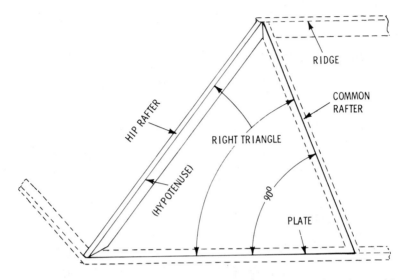

Fig. 17. The hip rafter is framed between the plate and the ridge and is the hypotenuse of a right-angle triangle whose other two sides are the adjacent common rafter and the intercepted portion of the plate.

The rise of the hip rafter is the same as that of the common rafter. The run of the hip rafter is the horizontal distance from the plumb line of its rise to the outside of the plate at the foot of the hip rafter. This run of the hip rafter is to the run of the common rafter as 17 is to 12. Therefore, for a ¹/₆ pitch, the common rafter run and rise are 12 and 4, respectively, while the hip rafter run and rise are 17 and 4, respectively.

For the top and bottom cuts of the common rafter, the figures are used that represent the common rafter run and rise, that is, 12 and 4 for a ¹/₆ pitch, 12 and 6 for a ¼ pitch, etc. However, for the top and bottom cuts of the hip rafter, use the figures 17 and 4, 17 and 6, etc., as the run and rise of the hip rafter. It must be remembered, however, that these figures will not be correct if the pitches on the two insides of the hip (or valley) are not the same.

Valley Rafters

The valley rafter is the hypotenuse of a right-angle triangle made by the common rafter with the ridge, as shown in Fig. 18.

This corresponds to the right-angle triangle made by the hip rafter with the common rafter and plate. Therefore, the rules for the lengths and cuts of valley rafters are the same as for hip rafters.

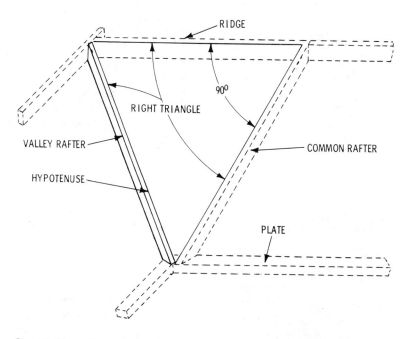

Fig. 18. The valley rafter is framed between the plate and the ridge and is the hypotenuse of a right-angle triangle whose other two sides are the adjacent common rafter and the intercepted portion of the ridge board.

Jack Rafters

These are usually spaced either 16 or 24 inches apart, and, since they lie equally spaced against the hip or valley, the second jack rafter must be twice as long as the first, the third three times as long as the first, and so on, as shown in Fig. 19. One reason for the 16- and 24-inch spacings on jack rafters is because of the roof sheathing; therefore, the rafters must be 16 or 24 inches apart so that the sheathing may be conveniently nailed to it.

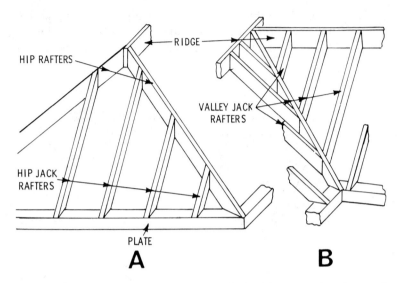

Fig. 19. Jack rafters. A, hip jack rafters, framed between the plate and hip rafters; B, valley jack rafters, framed between the ridge and the valley rafter.

Cripple Rafters

A cripple rafter is a jack rafter which touches neither the plate nor the ridge; it extends from the valley rafter to the hip rafters. The cripple-rafter length is that of the jack rafter plus the length necessary for its bottom cut, which is a plumb cut similar to the top cut. Top and bottom (plumb) cuts of cripples are the same as the top cut for jack rafters. The side cut at the hip and valley are the same as the side cut for jacks.

Finding Rafter Lengths Without the Aid of Tables

In the directions accompanying some framing squares and in some books, frequent mention is made of the figures 12, 13, and 17. Directions say, for example, that for common rafters "use figure 12 on the body and the rise of the roof on the tongue"; for hip or valley rafters, "use figure 17 on the body and the rise of the roof on the tongue" —but no explanation of how these fixed numbers are obtained is provided. The origin of such fixed numbers should be known to make the typical job easier to understand. They can be readily understood by referring to Fig. 20. In this illustration, let ABCD be a square whose sides are 24 inches long, and let abcdefgL be an inscribed octagon. Each side of the octagon (ab, bc, etc.) measures 10 inches; that is, LF = one-half side = 5 inches, and by construction, FM = 12 inches. Now, let FM represent the run of a common rafter. Then LM will be the run of an octagon rafter, and DM will be the run of a hip or valley rafter. The values for the run of octagon and hip or valley rafters (LM and DM, respectively) are obtained as follows:

$$LM = \sqrt{(FM)^2 + (LF)^2} = \sqrt{(12)^2 + (5)^2} = 13$$

$$DM = \sqrt{(FM)^2 + (DF)^2} = \sqrt{(12)^2 + (12)^2} = 16.97$$
$$or\ approximately\ 17$$

Example — What is the length of a common rafter having a 10-foot run and a ⅜ pitch?

For a 10-foot run,

$$the\ span = 2 \times 10 = 20\ feet$$

with ⅜ pitch,

$$rise = \frac{3}{8} \times 20 = 7.5\ feet$$

$$rise\ per\ foot\ run = \frac{rise \times 12}{run} = \frac{7.5 \times 12}{10} = 9\ inches$$

On the body of the square shown in Fig. 21, take 12 inches for 1 foot of run, and on the tongue, take 9 inches for the rise per foot of run. The diagonal, or distance between the points thus obtained, will be the length of the common rafter per foot of run with a ⅜ pitch. The distance FM measures 15 inches, or by calculation:

$$FM = \sqrt{(12)^2 + (9)^2} = 15\ inches$$

247

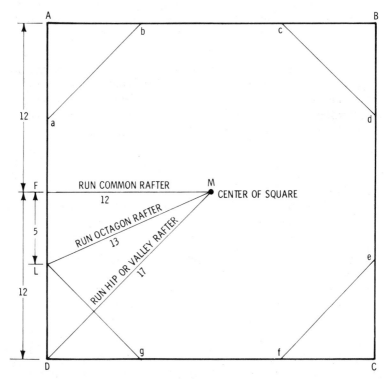

Fig. 20. A square and an inscribed octagon are used to illustrate the method of obtaining and using points 12, 13, and 17 in the application of a framing square to determine the length of rafters without the aid of rafter tables.

Since the length of run is 10 feet,

$$length\ of\ rafter\ =\ length\ of\ run\ \times\ length\ per\ foot$$
$$=\ 10\ \times\ ^{15}\!/_{12}$$
$$=\ \frac{150}{12}$$
$$=\ 12.5\ feet$$

The combination of figures 12 and 9 on the square, as shown in Fig. 21, not only gives the length of the rafter per foot of run, but, if the rule is considered as the rafter, the angles S and R for

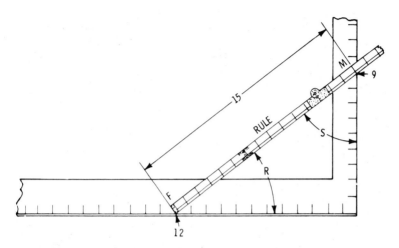

Fig. 21. A rule is placed on the square at points 12 and 9 to obtain the length of a common rafter per foot of run with a ⅜ pitch.

the top and bottom cuts are obtained. The points for making the top and bottom cuts are found by placing the square on the rafter so that a portion of one arm of the square represents the run and a portion of the other arm represents the rise. For the common rafter with a ⅜ pitch, these points are 12 and 9; the square is placed on the rafter as shown in Fig. 22.

Example—What length must an octagon rafter be to join a common rafter having a 10-foot run (as rafters MF and ML in Fig. 20)?

From Fig. 20, it is seen that the run per foot of an octagon rafter, as compared with a common rafter, is as 13 is to 12, and that the rise for a 13-inch run of an octagon rafter is the same as

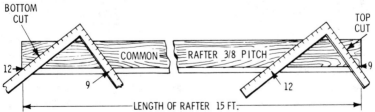

Fig. 22. The square is placed on the rafter at points 12 and 9, as shown, thereby giving the proper angles for the bottom and top cuts.

249

for the run of a 12-inch common rafter. Therefore, measure across from points 13 and 9 on the square, as MS in Fig. 23, which gives the length (15¾ inches) of an octagon rafter per foot of run of a common rafter. The length multiplied by the run of a common rafter gives the length of an octagon rafter; thus:

$$15\frac{3}{4} \times 10 = 157\frac{1}{2} \text{ inches} = 13 \text{ feet, } 1\frac{1}{2} \text{ inches}$$

Points 13 and 9 on the square (MS in Fig. 23) give the angles for the top and bottom cuts.

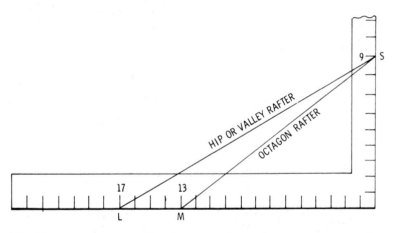

Fig. 23. Measurements using the square for octagon and hip or valley rafters, illustrating the use of points 13 and 17. Line MS (13,9) is the octagon rafter length per foot of run of a common rafter with a ⅜ pitch; line LS (17,9) is the hip or valley rafter length per foot of run of a common rafter with a ⅜ pitch.

Example — What length must a hip or valley rafter be to join a common rafter having a 10-foot run (as rafters MF and MD in Fig. 20)?

Fig. 20 shows that the run per foot of a hip or valley rafter, as compared with a common rafter, is as 17 is to 12, and that the rise per 17-inch run of a hip or valley rafter is the same as for a 12-inch run of a common rafter. Therefore, measure across from points 17 and 9 on the square, as LS in Fig. 23; this gives the length (19¼ inches) of the hip or valley rafter per foot of common rafter. This length, multiplied by the run of a common rafter, gives the length of the hip or valley rafter; thus:

$$19\frac{1}{4} \times 10 = 19\frac{1}{2} \text{ inches } = 16 \text{ feet, } \frac{1}{2} \text{ inch}$$

Points 17 and 9 n the square (LS in Fig. 23) give the angles for the top and bottom cuts.

Table 5 gives the points on the square of the top and bottom cuts of various rafters.

Table 5.   Square Points for Top and Bottom Cuts

	PITCH	1	11/12	5/6	3/4	2/3	7/12	1/2	5/12	1/3	1/4	1/6	1/12
Tongue	Common							12					
	Octagon							13					
	Hip or Valley							17					
	Body	24	22	20	18	16	14	12	10	8	6	4	2

RAFTER TABLES

The arrangement of these tables varies considerably with different makes of squares, not only in the way they are calculated but also in their positions on the square. On some squares, the rafter tables are found on the face of the body; on others, they are found on the back of the body. There are two general classes of rafter tables, grouped as follows:

1. Length of rafter per foot of run.
2. Total length of rafter.

Evidently, where the total length is given, there is no figuring to be done, but when the length is given per foot of run, the reading must be multiplied by the length of run to obtain the total length of the rafter. To illustrate these differences, directions for using several types of squares are given in the following sections. These differences relate to the common and hip or valley rafter tables.

Reading the Total Length of the Rafter

One popular type of square is selected as an example to show how rafter lengths may be read directly without any figuring.

The rafter tables on this particular square occupy both sides of the body instead of being combined in one table; the common rafter table is found on the back, and the hip, valley, and jack rafter tables are located on the face.

Common Rafter Table—The common rafter table, Fig. 24, includes the outside-edge graduations of the back of the square on both the body and the tongue; these graduations are in twelfths. The inch marks may represent inches or feet, and the twelfths marks may represent twelfths of an inch or twelfths of a foot (inches). The edge-graduation figures above the table represent the run of the rafter; under the proper figure on the line representing the pitch is found the rafter length required in the table. The pitch is represented by the figures at the left of the table under the word **PITCH**; thus:

12 Feet of Run						
Feet of Rise	4	6	8	10	12	15	18
Pitch	1/6	1/4	1/3	5/12	1/2	5/8	3/4

The length of a common rafter given in the common rafter table is from the top center of the ridge board to the outer edge of the plate. In actual practice, deduct one-half the thickness of the ridge board, and add for any eave projection beyond the plate.

Example—Find the length of a common rafter for a roof with a ¹/₆ pitch (rise = ¹/₆ the width of the building) and a run of 12 feet (found in the common rafter table, Fig. 24, the upper, or ¹/₆-pitch ruling).

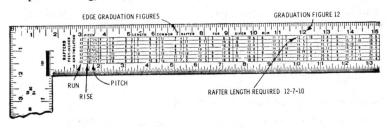

Fig. 24. The common rafter table.

Find the rafter length required under the graduation figure 12. This is found to be 12, 7, 10, which means 12 feet, $7^{10}/_{12}$ inches. If the run is 11 feet and the pitch is ½ (the rise = ½ the width of the building), then the rafter length will be 15, 6, 8, which means 15 feet $6^8/_{12}$ inches. If the run is 25 feet, add the rafter length for a run of 20 feet to the rafter length for a run of 5 feet. When the run is in inches, then in the rafter table read inches and twelfths instead of feet and inches. For instance, if, with a ½ pitch, the run is 12 feet 4 inches, add the rafter length of 4 inches to that of 12 feet as follows:

For a run of 12 feet, the rafter length is 16 feet, $11^8/_{12}$ inches.
For a run of 4 inches, the rafter length is $\qquad$ $5^8/_{12}$ inches.

Total—17 feet, $5^4/_{12}$ inches.

The run of 4 inches is found under the graduation 4 and is 5, 7, 11, which is approximately $5^8/_{12}$ inches. If the run was 4 feet, it would be read as 5 feet, $7^{11}/_{12}$ inches.

Hip Rafter Table—This table, as shown in Fig. 25, is located on the face of the body and is used in the same manner as the table for common rafters explained above. In the hip rafter table, the outside-edge-graduation figures represent the run of common rafters. The length of a rafter given in the table is from the top center of the ridge board to the outer edge of the plate. In actual practice, deduct one-half the thickness of the ridge board, and add for any eave projection beyond the plate. When using this table, find the figures on the line with the required pitch of the roof.

Under **PITCH,** the set of three columns of figures gives the

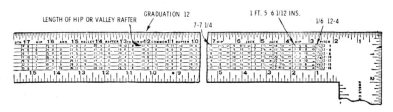

Fig. 25. The hip rafter table.

pitch. The seven pitches in common use are given, as for example $^1/_6$-12-4; this means that for a $^1/_6$ pitch, there is a 12-inch run per 4-inch rise.

Under **HIP,** the set of figures gives the length of the hip and valley rafter per foot of run of common rafter for each pitch, as 1 foot, $5^6/_{12}$ inches for a $^1/_6$ pitch.

Under **JACK** (16 inches on center), the set of figures gives the length of the shortest jack rafter, spaced 16 inches on center, which is also the difference in length of succeeding jack rafters.

Example—If the jack rafters are spaced 16 inches on center for a $^1/_6$-pitch roof, find the lengths of the jacks and cut bevels.

The jack top and bottom cuts (or plumb and heel cuts) are the same as for the common rafter. Take 12 on the tongue of the square; that is, mark on the 9½ sides, as shown in the illustration, which represents the rise per foot of the roof, or, if the pitch is given, take the figures in Table 5 that correspond to the given pitch. Thus, for a $^1/_6$ pitch, these points are 12 and 4. Fig. 26 shows the square on the jack in this position for marking top and bottom cuts.

Look along the line of $^1/_6$ pitch, in Fig. 27, under **JACK** (16-inch center), and find 16⅞, which is the length in inches of the shortest jack and is also the amount to be added for the second jack. Deduct one-half the thickness of the hip rafter, because the

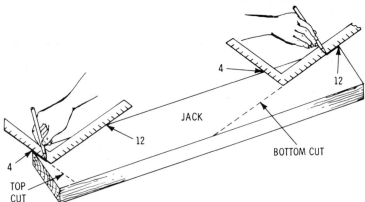

Fig. 26. The square is applied to a jack rafter for marking top and bottom cuts. The vertical and horizontal cuts for jack rafters are the same as for common rafters.

254

jack rafter lengths given in this table are to centers. Also, add for any projection beyond the outer edge of the plate.

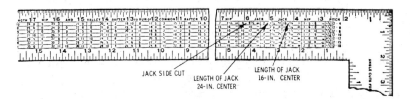

Fig. 27. A rafter table.

Look along the line of ¹⁄₆ pitch, in Fig. 27, under **JACK** (side cut), and find 9–9½ for a ¹⁄₆ pitch. These figures refer to the graduated scale on the edge of the square. To obtain the required bevel, take 9 on one arm and 9½ on the other, as shown in Fig. 28. It should be carefully noted that the last figure, or figure to

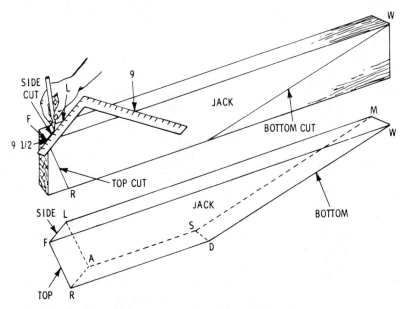

Fig. 28. Marking and cutting a jack rafter with the aid of the square. FR and DW are the marks for the top and bottom cuts, respectively. With the jack rafter cut as marked, L'A'R'F' represents the section cut at the top, and MSDW represents the section cut at the bottom

255

the right, gives the point on the marking side of the square; that is, mark on the 9½ sides, as shown in the illustration.

Under **JACK** (24 inches on center), the set of figures gives the length of the shortest jack rafter spaced 24 inches on center, which is also the difference in length of succeeding jack rafters. Deduct one-half the thickness of the hip or valley rafter, because the jack rafter lengths given in the table are to centers. Also, add for any projection beyond the plate.

Under **HIP,** the set of figures gives the side cut of the hip and valley rafters against the ridge board or deck, as 7–7¼ for a ⅙ pitch (mark on the 7¼ side).

To get the cut of the sheathing and shingles (whether hip or valley), reverse the figures under **HIP,** as 7¼–7 instead of 7–7¼. For the hip top and bottom cuts, take 17 on the body of the square, and, on the tongue, take the figure which represents the rise per foot of the roof.

Fig. 29 shows the marking and cut of the hip rafter, and Fig. 30 shows the rafter in position resting on the cap and the ridge. The section L′A′R′F′ resting on ridge is the same as L′A′R′F′ in Fig. 29.

Under **HIP AND VALLEY,** the set of figures gives the length of run of the hip or valley rafter for each pitch of the common rafter. For instance, for a roof with a ⅙ pitch under the figure 12 (representing the run of the common rafter, or one-half the width of the building), along the ⅙-pitch line of figures find 17, 5, 3, which means 17 feet, $5^3/_{12}$ inches, which is the length of the hip or valley rafter. Deduct one-half the thickness of the ridge board, and add for eave overhang beyond the plate, which is the length of the hip or valley rafter required for a roof with a ⅙ pitch and a common rafter run of 12 feet.

Example — Find the length of the hip rafter for a building that has a 24-foot span and a ⅙ pitch (a 4-inch rise per foot of run).

In the hip rafter table (Fig. 25) along the line of figures for ⅙ pitch and under the graduation figure 12 (representing one-half the span, or the run of the common rafter), find 17, 5, 3, which means 17 feet, $5^3/_{12}$ inches; this is the required length of the hip or valley rafter. Deduct one-half the thickness of the ridge board, and add for any overhang required beyond the plate.

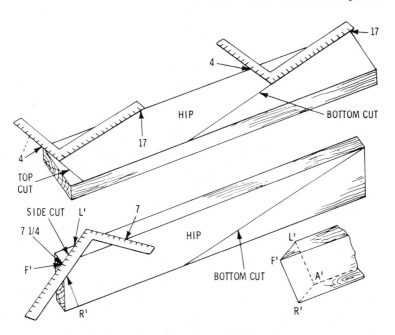

Fig. 29. The square, as applied to hip rafters, for marking top, bottom, and side cuts. Note that the number 17 on the body is used for hip rafters. Section L'A'R'F' shows the bevel required for the ridge.

For the top and bottom cuts of the hip or valley rafter, take 17 on the body of the square and 4 (the rise of the roof per foot) on the tongue. The mark on the 17 side gives the bottom cut; the mark on the 4 side gives the top cut.

For the side cut of the hip or valley rafter against the ridge board, look in the set of figures for the side cut in the table (Fig. 25) under **HIP** along the line for $\frac{1}{6}$ pitch, and find the figure 7–7¼. Use 7 on one arm of the square and 7¼ on the other; mark on the 7¼ arm for the side cut.

Reading Length of Rafter per Foot of Run

There are many methods used by carpenters for determining the lengths of rafters, but probably the most dependably accurate method is the "length-per-foot-of-run" method. Since many, perhaps most, of the better rafter-framing squares now have tables on their blades giving the necessary figures, they

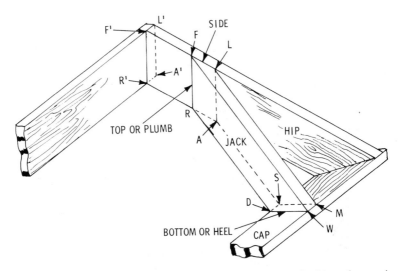

Fig. 30. The jack rafter in position on the roof between the hip rafter and the cap.

may almost be considered as standard. The tables may not be arranged in the same manner on all these squares, and on some they may be more complete than on others.

Under the heading **LENGTH COMMON RAFTERS PER FOOT RUN** in Fig. 31 will be found numbers, usually from 3 to 20. With each number is a figure in inches and decimal hundredths. The integers represent the rises of the rafters per foot of run, and the inches and decimals represent the lengths of the rafters per foot of run. As an example of the use of these tables, take a building 28 feet, 2 inches wide, thereby making the run of the rafters 14 feet (allowing for a 2-inch ridge). Let the desired

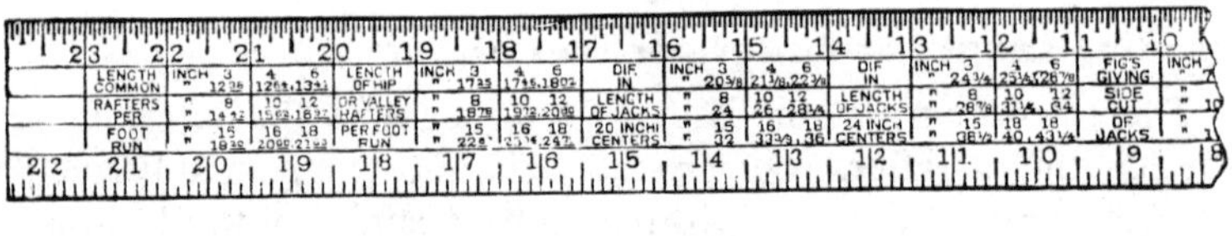

Fig. 31. The length-per-foot-run tables on one type of rafter framing square.

pitch be 4 inches per foot. Under the number 4 on the square will be found the length per foot of run — 12.64 inches. The calculation for the length of the rafter is as follows:

$$12.64 \times 14 = 176.96 \text{ inches}$$

$$\frac{176.96}{12} = 14.75 \text{ feet, or 14 feet, 9 inches}$$

If the run is in feet and inches, it is most convenient to reduce the inches to the decimal parts of a foot, according to the following table:

1 inch	equals	0.083	foot
2 inches	equals	0.167	foot
3 inches	equals	0.250	foot
4 inches	equals	0.333	foot
5 inches	equals	0.417	foot
6 inches	equals	0.500	foot
7 inches	equals	0.583	foot
8 inches	equals	0.667	foot
9 inches	equals	0.750	foot
10 inches	equals	0.833	foot
11 inches	equals	0.917	foot

Example — Find the length of a hip or valley rafter having an 8-inch rise per foot on a 20-foot building with the run of the common rafters measuring 10 feet.

Look for **LENGTH OF HIP OR VALLEY RAFTERS PER FOOT RUN** (Fig. 31), and read under the 8-inch rise the figure 18.76. This is the calculation:

$$18.76 \times 10 = 187.6 \text{ inches}$$

$$\frac{187.6}{12} = 15 \text{ feet, 7.6 inches}$$

One edge of all good steel squares is divided into tenths of inches, so this length may be measured off directly on the rafter pattern with the steel square.

Example—Find the difference in lengths of jack rafters on a roof with an 8-inch rise per foot and with a spacing of 16 inches on centers.

Under **DIFFERENCE IN LENGTH OF JACKS** (16-inch centers) on the square find the figure 19.23 below the figure 8 (rise per foot of run). This is the length of the first jack rafter, and the length of each succeeding jack will be 19.23 inches greater— 38.46 inches, 57.69 inches, 76.92 inches, etc.

Example—Find the side cuts of jacks on a square similar to the one shown in Fig. 32.

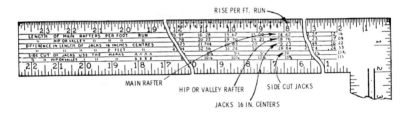

Fig. 32. Typical rafter tables.

The fifth line is marked **SIDE CUT OF JACKS USE THE MARKS** $\wedge$. If the rise is 8 inches per foot, find the figure 10 under figure 8 in the upper line. The proper side cut will then be 10 × 12, cut on 12. The side cuts for hip or valley rafters are found in the sixth line; for the 8 × 12 roof, it is 10⅞ × 12, cut on 12.

No discussion of rafter framing is complete without an explanation of one of the oldest and most useful, though probably not the most accurate, methods of laying out a rafter with a steel square. Any square may be used if it has legible inch marks representing the desired pitch. It is the same method used for the layout of stairs. Fig. 33 shows the layout for a rafter with a 9-foot run that has a pitch of 7 inches × 12 inches, making the rise of the rafter 5 feet 3 inches. The steel square is applied nine times; carefully mark each application, preferably with a knife. A hip rafter is laid out in exactly the same manner by using 17 instead of 12 in the run and applying the square nine times as was done for the common rafter. For short rafters, this is probably the least time-consuming of any method.

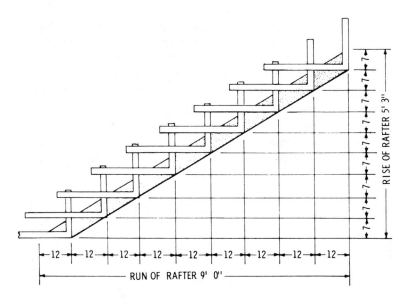

Fig. 33. The method of stepping off a rafter with a square; the square is applied in consecutive steps, hence the name of the method.

Table of Octagon Rafters

The complete framing square is provided with a table for cutting octagon rafters, as shown in Fig. 34. In this table, the first line of figures from the top gives the length of octagon hip rafters per foot of run. The second line of figures gives the length of jack rafters spaced 1 foot from the octagon hip. The third line of figures gives the reference to the graduated edge that will give the side cut for octagon hip rafters. The fourth line of figures gives the reference to the graduated edge that will give the side

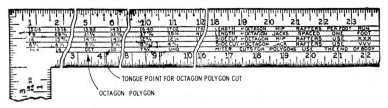

Fig. 34. Typical octagon rafter tables.

261

cuts for jack rafters. The tables are used in a manner similar to that used for the regular rafter tables just described and, therefore, need no further explanation. The last line, or bottom row of figures, gives the bevel of intersecting lines of various regular polygons. At the right end of the body on the bottom line can be read **MITER CUTS FOR POLYGONS—USE END OF BODY.**

Example—Find the angle cut for an octagon.

For a figure of 8 sides, look to the right of the word **OCT** in the last line of figures, and find 10. This is the tongue reading; the end of the body is the other point, as shown in Fig. 35.

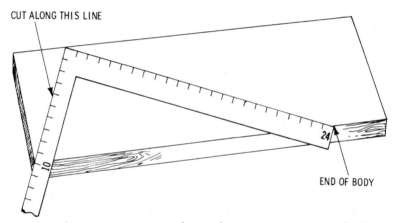

CUT ALONG THIS LINE

24

END OF BODY

Fig. 35. The square in position for marking an octagon cut; it is set to point 10 on the tongue and to point 24 on the body.

TABLE OF ANGLE CUTS FOR POLYGONS

This table is usually found on the face of the tongue. It gives the setting points at which the square should be placed to mark cuts for common polygons that have from 5 to 12 sides.

Example—Find the bevel cuts for an octagon.

On the face of the tongue (Fig. 36), look along the line marked **ANGLE CUTS FOR POLYGONS,** and find the reading "8 sides 18–7½." This means that the square must be placed at 18 on one arm and at 7½ on the other to obtain the octagon cut, as shown in Fig. 37.

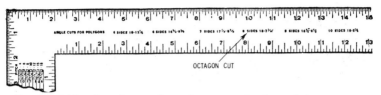

Fig. 36. Table of angle cuts for polygons on the face of the square.

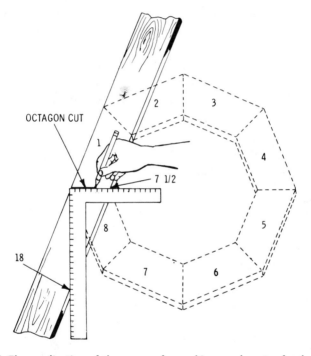

Fig. 37. The application of the square for making angle cuts of polygons. The square is shown set to points 18 and 7½. When constructing an 8-sided figure, such as an octagon cap, the last figure in the reading is the setting for making the side. Mark as shown. Cut eight pieces to equal length, with this angle cut at each end of each piece. The pieces will fit together to make an 8-sided figure, as shown by the dotted lines.

TABLE OF BRACE MEASURE

This table on the square, shown in Fig. 38, is located along the center of the back of the tongue and gives the length of common braces.

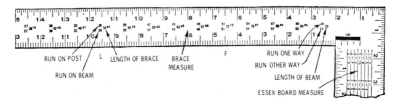

Fig. 38. Table of brace measure on the back of the square.

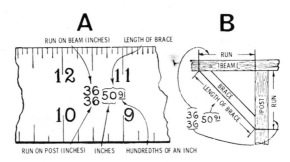

Fig. 39. A portion of the brace-measure table, with an explanation of the various figures, is shown in A. The brace in position, illustrating the measurements of the brace-measure table, is shown in B.

Example—If the run is 36 inches on the post and 36 inches on the beam, what is the length of the brace?

In the brace table along the central portion of the back of the tongue (Figs. 38 and 39), look at L for

$$36$$
$$50.91$$
$$36$$

This reading means that for a run of 36 inches on the post and 36 inches on the beam, the length of the beam is 50.91 inches.

At the end of the table (at F near the body) will be found the reading

$$18$$
$$30$$
$$24$$

This means that where the run is 18 inches one way and 24 inches the other, the length of the brace is 30 inches.

The best way to find the length of the brace for the runs not given on the square is to multiply the length of the run by 1.4142 feet (when the run is given in feet) or by 16.97 inches (when the run is given in inches). This rule applies only when both runs are the same.

OCTAGON TABLE OR EIGHT-SQUARE SCALE

This table on the square is usually located along the middle of the tongue face and is used for laying off lines to cut an eight-square or octagon-shaped piece of timber from a square timber.

In Fig. 40, let ABCD represent the end section, or butt, of a square piece of 6″ × 6″ timber. Through the center draw the lines AB and CD parallel with the sides and at right angles to each other. With dividers take as many squares (6) from the scale as there are inches in width of the piece of timber, and lay off this square on either side of the point A, such as Aa and Ah; lay off in the same way the same spaces from the point B, as Bd and Be; also lay off Cb, Cc, Df, and Dg. Then draw the lines ab, cd, ef, and gh. Cut off at the edges to lines ab, cd, ef, and gh, thus obtaining the octagon or 8-sided piece.

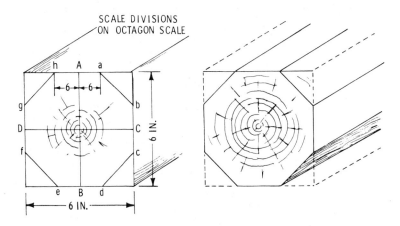

Fig. 40. A square timber and its appearance after it has been cut to an octagon shape, which shows the application of the octagon scale.

265

ESSEX BOARD MEASURE TABLE

This table is illustrated in Fig. 41 and normally appears on the back of the tongue on the square. To use the table, the inch graduations on the outer edge of the square are used in combination with the values along the five parallel lines. After measuring the length and width of the board, look under the 12-inch mark for the width in inches. Then follow the line on which this width is stamped toward either end until the inch mark is reached on the edge of the square where the number corresponds to the length of the board in feet. The number found under that inch mark will be the length of the board in feet and inches. The first number is feet, and the second is inches.

Instead of a dash between the foot and inch numbers, some squares have the inch division continued across the several parallel lines of the scale appearing on one side of the vertical inch division lines and inches on the other.

Example — How many feet Essex board measure in a board 11 inches wide, 10 feet long, and 1 inch thick? 3 inches thick?

Under the 12-inch mark on the outer edge of the square (Fig. 41) find 11, which represents the width of the board in inches. Then follow on that line to the 10-inch mark (representing the length of the board in feet), and find on a line 9–2, which means that the board contains 9 feet 2 inches board measure for a thickness of 1 inch. If the thickness were 3 inches, then the board would contain 9 feet 2 inches × 3, or 27 feet 6 inches B.M.

SUMMARY

The success of any workshop operation depends on having a good array of tools and having a knowledge of their operations.

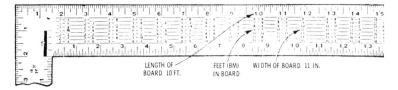

Fig. 41. Table of Essex board measure on the back of the square.

On most construction work, especially in house framing, the steel square is invaluable for accurate measuring and for determining angles.

The square most generally used has an 18-inch tongue and a 24-inch body. The body is generally 2 inches wide, and the tongue is 1½ inches wide, varying in thickness from $3/16$ to $3/32$ of an inch. The various markings on the square are tables and scales or graduations.

Since the tables on the square relate mostly to problems encountered in cutting lumber for roof-frame work, it is necessary to know roof construction and the names of various rafters. These names are rise, run, rise per foot run, hip and valley rafters, jack and cripple rafters, common rafters, ridge and plate. An example of these various members is shown in Fig. 12 of this chapter.

The rafter tables vary considerably with different makes of squares, not only in the way they are calculated but also in their positions on the square. Some tables are found on the face of the body, and others are on the back of the body. The two general classes of rafter tables found on squares are length of rafter per foot of run, and total length of rafter.

REVIEW QUESTIONS

1. It is called a steel square, but what is the correct name for this tool?
2. What type of tables are found on the body of the square?
3. Name the various types of roof rafters.
4. What is rise per foot run of a roof?
5. What is rafter pitch?

CHAPTER 21

Joints and Joinery

The most challenging job for anyone who works with wood is making joints. The problem is to cut two pieces of wood so that they meet without gaps—or with very little space—so the joint can do the job it is intended to do. It is a job that requires precision and patience, and the ability to make good joints is the hallmark of woodworkers who have mastered their trade.

Before modern glues and clamps, many joints relied on various wedges and keys and stresses to hold the joint together. Today, though, with tenacious glues and proper clamping procedures and equipment—plus the availability of special fasteners as well as nails and screws and bolts—the job can be done with relative ease. Indeed, most carpenters get by with just a few joints, such as butt, rabbet, dado, and miter.

Following is a presentation of the main joints used today plus a consideration of some joints that the average carpenter or handyman will rarely see but should know about as part of a well-rounded corpus of knowledge. For craftsmen interested in working on antique furniture or other old items—such as old boats—a knowledge of these joints can be important, even crucial, to do the job required.

There is no use trying to classify all types of wood joints because their number and descriptions are infinite, but many of them may be placed under these headings:

1. Straight butt.
2. Dowel.
3. Square, butted, or mitered corners.

4. Dado.
5. Scarf.
6. Mortise and tenon.
7. Dovetail.
8. Wedge.
9. Tongue and groove.
10. Rabbet.

PLAIN EDGE AND BUTTED JOINTS

The plain edge joint is a joint between the edges of boards where the side of one piece is placed against the side of another, whereas the butt joint is a joint in which the square end of one member is placed against the square end of another.

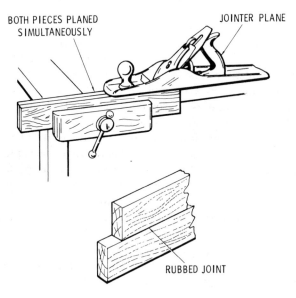

Fig. 1. The method of planing both edges together to obtain a straight-butt side joint. This requires a great deal of skill in planing, and it is necessary that the plane be straight on the edge and carefully sharpened and adjusted. After planing, the edges are glued and rubbed together to bond securely.

Straight Plain Edge Joint

This type of joint is more or less readily made on a power jointer. The plain edge joint has many uses and is commonly used to build up wide boards for panels, shelves, etc., from narrower pieces. For boat planking, the boards are often curved and slightly beveled so that the joint is left open to be calked later. Such curved joints must be fitted one edge to the other. In furniture, cabinet, and other fine finish work, the edges are usually glued.

To make a glued edge joint, square and straighten the edges carefully with planes. Test the edges often with the try square to assure squareness. To join the edges, polyvinyl (white) glues can be used. They dry quickly, are nonstaining, and no special equipment is needed for their application. White glue is not waterproof. If the work is to be used outside, use resorcinol glues; they are waterproof, although they are red in color and can stain. They can be painted over readily.

All modern glues will function at room temperature. If several boards are to be joined edge to edge, as indicated in Fig. 2, at least three clamps will usually be necessary—one on one side and two on the other side—to prevent buckling. Take care to make the edges true even when gluing, or it may unnecessarily require considerable scraping to make the joint flush and smooth.

Dowel Joints

It is usually not necessary to dowel a well-fitting glued edge joint, but it is sometimes done to facilitate assembly; the dowels used are usually quite short. For a butt joint into side wood,

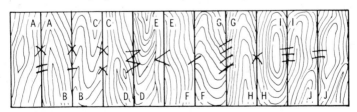

Fig. 2. Narrow boards can be jointed and placed together by using a marking system so that the same edges will come together when assembling them.

271

they are a satisfactory substitute for mortise and tenon joints and are considerably easier to make. When making heavy screen frames, storm sash, etc., dowel joints are satisfactory if they are glued together with waterproof glue. Fig. 3 shows the assembly of a typical dowel joint.

The holes must be accurately marked and bored; if these precautions are not taken, the holes will not be in perfect alignment and it will be impossible to assemble the joint, or, when assembled, the pieces will not be in their proper alignment. Jigs designed to hold the bit in alignment are obtainable from several major tool companies, and these devices are a great help when a great amount of doweling is to be done.

The method of making dowel joints without a jig is shown in Fig. 4. Dowel rods made of several different types of hardwoods are obtainable; some of them have shallow spiral grooves around them to assist carrying the glue into the hole.

Square Corner Joint

The two members of a corner joint are joined at right angles, the end of one butting against the side of the other. When making a corner joint, saw to the squared line with a backsaw and finish with a block plane to fit. The work should be frequently tested with a try square, both lengthwise and across the joint. The method of marking this type of joint is shown in Fig. 5. The joint may be fastened together with nails or screws. When fas-

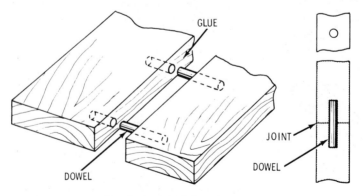

Fig. 3. A typical dowel pin joint.

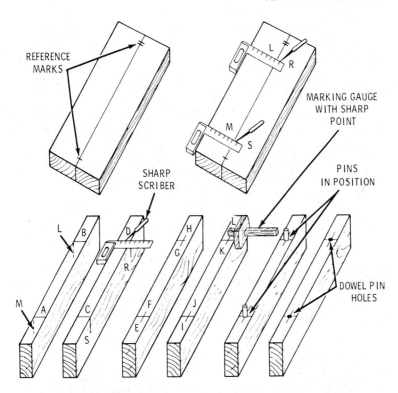

REFERENCE MARKS

MARKING GAUGE WITH SHARP POINT

SHARP SCRIBER

PINS IN POSITION

DOWEL PIN HOLES

Fig. 4. The method of making dowel joints. After making reference marks on the two boards, scribe lines A, B, C, and D. Set the marking gauge to half the thickness of the boards, and scribe lines EF, GH, IJ, and KL. Bore a hole at the intersection of each of these lines; the holes should be just less than half the thickness of the boards. The dowels should fit tightly in these holes.

tening the joint, the pieces should be firmly held in position at a 90° angle by a vise or by some other suitable means.

Miter Joint

A miter joint is used mostly in making picture frames. To properly make a miter joint, a picture-frame vise should be used when fastening the pieces together instead of the makeshift method of offset nailing. In fact, a picture-framing shop, to be worthy of the name, should be provided with a picture-frame vise, one type of which is shown in Fig. 6.

273

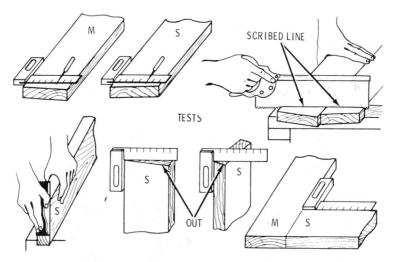

Fig. 5. The method of making a corner joint. After squaring and sawing the edges (M and S), plane the joint surface of one board (S), and test the edges with the square until a perfect right-angle fit is obtained.

Fig. 6. A typical picture frame vise. With this tool, any frame can be held in the proper position for nailing.

When cutting the 45° miter, use a miter box. After sawing, dress and fit the ends with a block plane. There are two ways to nail a miter joint—the correct way with a picture-frame vise and the wrong way with an ordinary vise. Where considerable work is to be done, a combined miter box and vise is desirable; one of these is shown in Fig. 7.

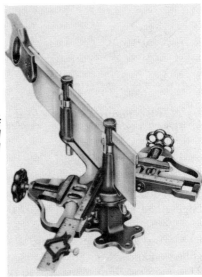

Fig. 7. A typical miter machine. With this device, any type of miter joint can be cut, glued, and nailed to make tight, close-fitting corners.

The methods of mitering corners shown in Fig. 8 are used to a great extent when constructing small drawers for merchandise cabinets, such as those used in drug stores. The joinery shown in Fig. 8A is not particularly effective. A much stronger joint may be made by sawing the groove for the feather straight across the corner almost through; then glue in strong hardwood feathers, with their ends cut off and smoothed flush. The method shown in Fig. 8B is not too efficient and is difficult to make, since the outside corner of the groove is often chipped in construction. Making the joint can be made easier by sawing the grooves on a band saw, or on a jigsaw if there are many to make, and then driving a small patented metal feather with sharp turned edges and a slight taper in from each edge. The feathers draw the joint tight, hold well, and no blocking is necessary.

275

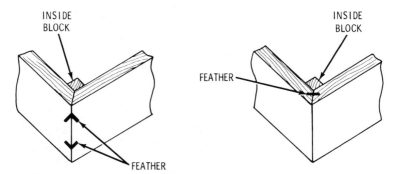

A. Miter joint with two outer-edge feathers.

B. Miter joint with end feathers.

Fig. 8. Miter joints reinforced by feathers. These feathers are kept in place by glue; the joint may also be reinforced by an inside block, as shown.

Splined Joint

The form of joint shown in Fig. 9 is called a splined joint, or sometimes a slip-tongue joint. In the shop, it is often used for edge-glued joints, since it holds the members in alignment when clamped.

A groove is made in each of the pieces to be joined, and a spline, made as a separate piece, is inserted in both grooves. The main reason for the use of a splined joint is that when two pieces of softwood are joined, a hardwood spline (which should

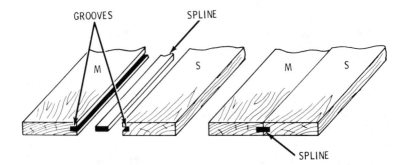

Fig. 9. The component parts and assembly of a splined joint. The spline fits into the grooves in M and S.

be cut across the grain) will make the joint less likely to snap than if a tongue were cut in the softwood lengthwise with the grain.

Splice Joints

A splice joint can be used to join pieces of wood end to end; they are joined by "fish plates" placed on each side and are secured by cross bolts, or nails, as shown in Fig. 10. These fish plates may be made either of wood or of iron, and they may have plain or projecting ends.

The plain type, shown in Fig. 10A, is normally suitable when the form of stress is compression only; however, if the joint is properly made, it will withstand either tension or compression. If the joint is to be subjected to tension, the fish plates (either wood or iron) should be anchored to the main members by keys or projections, as shown in Fig. 10C and D.

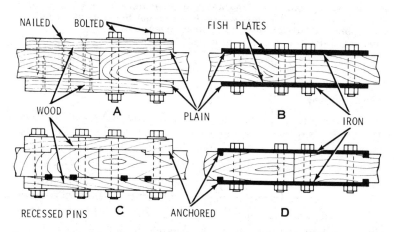

Fig. 10. Various splice, or fish, joints; A, plain joint with wooden fish plates; B, plain joint with iron fish plates; C, wood plates anchored on the end; D, iron plates anchored on the end.

LAP JOINTS

In the various joints grouped under this classification, one of the pieces to be joined laps over, or into, the other, hence the name lap joint. Some typical lap joints are shown in Fig. 11.

277

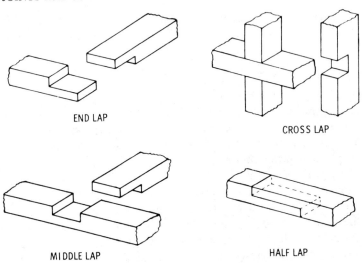

END LAP

CROSS LAP

MIDDLE LAP

HALF LAP

Fig. 11. Typical lap joints. The overlapping feature furnishes a greater holding area in the joint and is therefore stronger than any of the butt or plain joints. A half-lap joint, sometimes called a scarf joint, is made by tapering or notching the sides or ends of two members so that they overlap to form one continuous piece without an increase in thickness. The joints are usually fastened with plates, screws, or nails and are strengthened with glue.

Rabbet Joint

A rabbet joint is cut across the edge or end of a piece of stock. The joint is cut to a depth of about ½ the thickness of the material. The rabbet joint is a common joint in the construction of cabinets and furniture because it allows pieces to be joined so that no seam shows. On a typical cabinet, for example, the recess, or L-shaped section, of the joint can be cut out of the end of each side and the back of the cabinet simply fitted in between the sides and fastened in place. Anyone looking at the cabinet from the sides won't see any seam.

Rabbet joints are also used in the construction of drawers, and in the making of boxes of various kinds. A rabbet joint can be made with a wide variety of tools.

Dado Joint

A dado joint is a groove cut across the grain which will receive the butt end of a piece of stock. The dado is cut to the width of the stock that will fit in it and to a depth of ½ the thickness of the material. The joint is a common one in construction and is used for the installation of shelves, stairs, and kitchen cabinets.

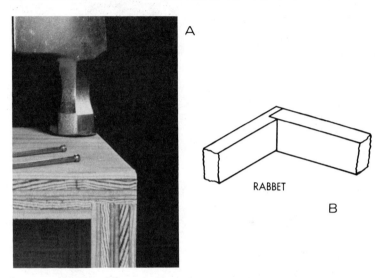

Fig. 12. A: Rabbet joint. An L-shaped section is cut out of the end of one board and the other board is slipped into the recess, then nailed, screwed, glued, or otherwise attached. B: Another view of the rabbet joint. 12A Courtesy of The American Plywood Assn.

Scarf Joints

By definition, a scarf joint is made by cutting away the ends of two pieces of timber and by chamfering, halving, notching, or sloping, making them fit each other without increasing the thickness at the splice. They may be held in place by gluing, bolting, plating, or strapping.

There are various forms of scarf joints, and they may be classified according to the nature of the stresses which they are designed to resist, as:

1. Compression.
2. Tension.
3. Bending.
4. Compression and tension.
5. Tension and bending.

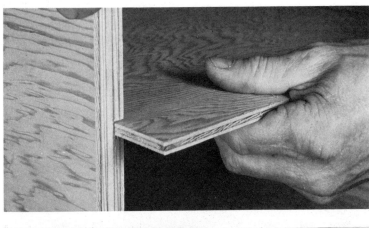

Fig. 13. A: dado joint. This is basically a groove cut across the grain, the groove being the width of the material. The dado is a very popular joint for shelves. B: Another view of dado joint. Here, a Surform tool is being used to clean out dadoes made on the upright portion of a shelf unit. 13A Courtesy of The American Plywood Assn. 13B Courtesy of Stanley.

Compression Scarf Joint—This is the simplest form of scarf joint. As usually made, one-half of the wood is cut away from the end of each piece for a distance equal to the lap, as shown in

Fig. 14A; this process is called "halving." The length of the lap should be five to six times the thickness of the timber. Mitered ends, as in Fig. 14B, are better than square ends, where nails or screws are depended on to fasten the joint. For extra heavy-duty joints, iron fish plates are sometimes provided, thereby greatly strengthening the joint, as shown in Fig. 14C; when these are used, mitered ends are not necessary.

Tension Scarf Joint—There used to be various methods of "locking" joints to resist tension, such as by means of keys, wedges, or so-called keys or fish plates with fingers, etc., as shown in Figs. 15 and 16. Today, you will still see fish plates used, but the keys and wedges are not used in modern joinery.

Bending Scarf Joint—When a beam is acted on by a transverse, or bending stress, the side on which the bending force is applied is subjected to a compression stress, and the opposite

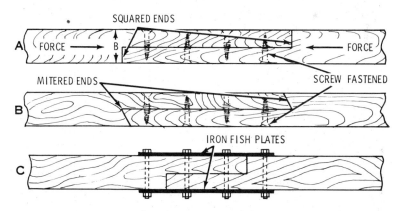

Fig. 14. Compression scarf joints; A, plain, square ends; B, plain, mitered ends; C, plain, square ends, reinforced with iron fish plates.

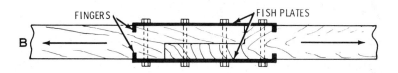

Fig. 15. Square ends bolted and reinforced with iron fish plates; tension stress caused by fingers on the fish plates.

281

KEY   WEDGES

Fig. 16. Key and wedges.

side is subjected to a tension stress. Thus, in Fig. 18A, the upper side is in compression, and the lower side is in tension. At L, the end of the joint may be square, but at F, it should be mitered. If this end were square (as at F', Fig. 18B), the portion of the lap M between the bolt and F' would be rendered useless to resist the bending force.

When designing a bending scarf joint, it is important that the thickness at the mitered end be ample, otherwise the strain applied at that point might split the support. Gluing normally helps prevent such stresses from developing.

Mortise-and-Tenon Joints

A mortise is defined as a space hollowed out in a member to receive a tenon, and a tenon is defined as a projection, usually

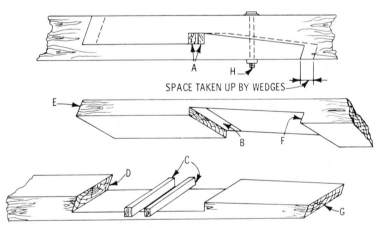

SPACE TAKEN UP BY WEDGES

Fig. 17. A scarf joint with a notch and a mitered half lap; the ends are also mitered, illustrating the location and effect of the wedges. The two pieces are joined together with the wedges (A) driven home and cut off. The dotted lines represent the amount of space closed when the pieces are drawn into place by the wedges. Since the cut is mitered at D, E, F, and G, these boards will form a rigid joint, which is often strengthened by a bolt through each section (H).

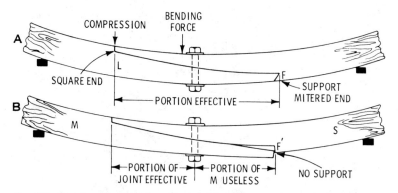

Fig. 18. Bending scarf joints. One end of the joint should be mitered to provide adequate support for the various stresses applied to the joint.

with a rectangular cross section, at the end of a piece of member that is to be inserted into a socket, or mortise, in another timber to make a joint.

Mortise-and-tenon joints are frequently called simply tenon joints. The operation of making mortise-and-tenon joints is also termed tenoning, which also implies mortising.

There are many different mortise-and-tenon joints, as illustrated in Figs. 20-26, and they may be classified with respect to:

1. Shape of the mortise.
2. Position of the tenon.
3. Degree in which tenon projects into mortised member.
4. Degree of mortise housing.
5. Number of tenons.
6. Shape of tenon shoulders.
7. Method of fastening the tenon.

The mortise and tenon must exactly correspond in size; that is, the tenon must accurately fit into the mortise. The position of the tenon is usually at the center of the member, but sometimes it is located at the side, depending, except in special cases, on the degree of housing. The tenon may project partly into, or through, the mortised member. When the tenon and mortise do not extend through the mortised member, the joint is called a stub tenon. This form of tenon is used for jointing the framework

283

Fig. 19. Most people think of the mortise-and-tenon joint as a square part fitting into a round hole, but mortise-and-tenon joints are popular when the tenon is round and so is the mortise, as is the case on this chair.
Courtesy of United Gilsonite.

of partitions and is also employed in work where the joint will not be subjected to any tension.

The term "degree of housing" signifies the degree in which the tenon is covered by the mortise, that is, the number of sides of the mortise. The number of tenons depends on the shape of the members, whether they are square or rectangular, with considerable width and little thickness, etc. The tenon shoulders are usually at right angles with the tenon as they are when the two

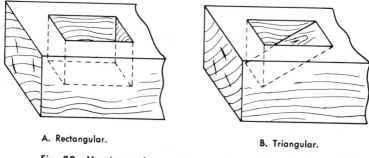

A. Rectangular. B. Triangular.

Fig. 20. Mortise-and-tenon joints—shape of the mortise.

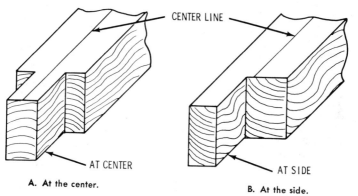

CENTER LINE

AT CENTER AT SIDE

A. At the center. B. At the side.

Fig. 21. Mortise-and-tenon joints—position of the tenon.

members are joined at right angles, but they may be mitered to some smaller angle, such as 60° or 45°, as in the case of a brace.

There are several ways of fastening mortise-and-tenon joints, such as with pins or wedges. When making a mortise and tenon joint, the work is first laid out to given dimensions, as shown in Fig. 27.

Cutting the Mortise—Select a chisel that is as near to the width of the mortise as possible. This chisel, especially for large work, should be a framing or mortise chisel. Bore a hole the same size as the width of the mortise at the middle point. If the mortise is for a through tenon, bore halfway through from each side. In the case of a large mortise, most of the wood may be

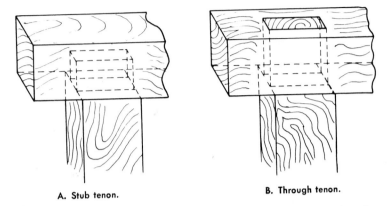

A. Stub tenon.   B. Through tenon.

Fig. 22. Mortise-and-tenon joints—degree in which the tenon projects into the mortised timber.

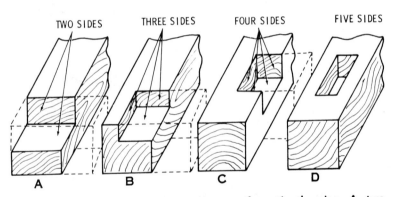

TWO SIDES   THREE SIDES   FOUR SIDES   FIVE SIDES

A   B   C   D

Fig. 23. Mortise-and-tenon joints—degree of mortise housing; A, two sides; B, three sides; C, four sides; D, five sides.

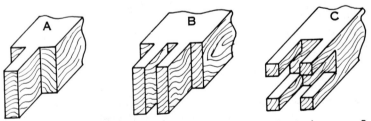

A   B   C

Fig. 24. Mortise-and-tenon joint—number of tenons; A, single tenon; B, double tenon; C, multitenon. The double and multitenon have not been used for many years.

286

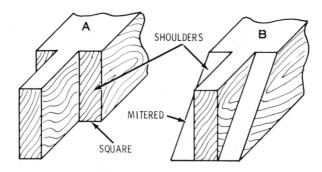

Fig. 25. Mortise-and-tenon joints—shape of the tenon shoulders; A, square; B, mitered.

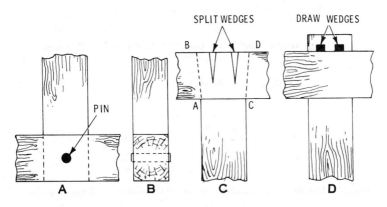

Fig. 26. Shown are former methods of fastening the tenon; A, side view, tenon secured by a pin; B, front view, tenon secured by a pin; C, tenon secured by internal, or split, wedges; sides AB and CD are tapered, thus securely wedging the tenon into the mortise; D, tenon secured by external, or draw, wedges, which are driven into rectangular holes beyond the mortise. Today, strong glue and proper clamping usually do the job.

removed by boring several holes, as shown in Fig. 28. When cutting out a small mortise with a narrow chisel, work from the hole in the center to each end of the mortise, holding the chisel firmly at right angles with the grain of the wood. At the ends of the mortise, the chisel must be held in a vertical position, as shown in Fig. 29B, with the flat side facing the end of the mortise.

287

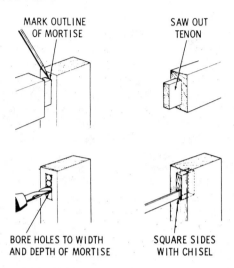

Fig. 27. *The method of laying out and making a small mortise-and-tenon joint.*

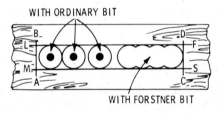

Fig. 28. *The method of boring holes when making a large mortise.*

Always loosen the chisel by a backward movement of the handle; a movement in the opposite direction would damage the ends of the mortise. Never make a chisel cut parallel with the grain because the wood at the side of the mortise may split. When cutting a through mortise, cut only halfway through on one side, and finish the cut from the other side. After cutting, test the sides of the mortise by using a try square, as shown in Fig. 30; this procedure will check the accuracy with which the work was laid out.

Cutting the Tenon—A backsaw is used for cutting out the wood on each side of the tenon, and, if necessary, a finishing cut

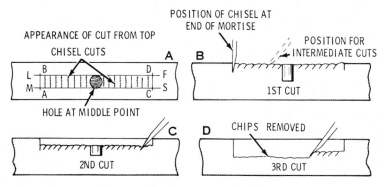

Fig. 29. The method of cutting a small mortise. After laying out the mortise, bore a hole at the center (A) and work toward each end with a chisel. The chisel cuts should always be made across the grain.

Fig. 30. Test the end with a square after cutting the mortise.

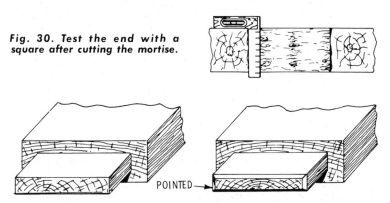

Fig. 31. Appearance of the tenon before and after pointing.

may be taken with a chisel. After the wood has been cut away, the tenon should be pointed by chiseling all four sides.

Fig. 31 shows the appearance of the tenon before and after the pointing operation; if this operation were omitted, a tight-fitting tenon would be difficult to start into the mortise and could splinter the sides of the mortise when driven through. Do not cut off the point until the tenon is finally in place and the pin is driven home.

Draw Boring—The term "draw boring" signifies the method of locating holes in the mortise and tenon that are eccentric with each other so that when the pin is driven in, it will "draw" the tenon into the mortise, thereby forcing the tenon shoulders tightly against the mortised member. The holes may be located either by accurately laying out the center, as shown in Fig. 32C, or by boring the mortise and finding the center for the tenon hole, as in Fig. 32D. Considerable experience is necessary to properly locate the tenon hole. If too much offset is given, an undue strain will be brought to bear on the joint; this strain is frequently sufficient to split the joint. It is much better to accurately lay out the work and make a tightly fitting pin than to depend on draw boring.

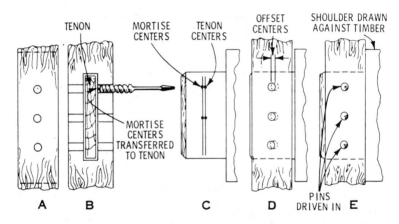

Fig. 32. The method of transferring pin centers from the mortise holes to the tenon by draw boring. When laying out the tenon-hole centers, make the offset toward the tenon shoulder.

Dovetail Joints

A dovetail joint may be defined as a partially housed tapered mortise-and-tenon joint, the tapered form of mortise and tenon forming a lock which securely holds the parts together. The word "dovetail" is used to describe the way the tenon expands toward the tip and resembles the fan-like form of the tail of a dove. The various forms of dovetail points, some of which are shown in Fig. 33, may be classed as:

1. Common.

2. Compound.
3. Lap, or half-blind.
4. Mortise, or blind.

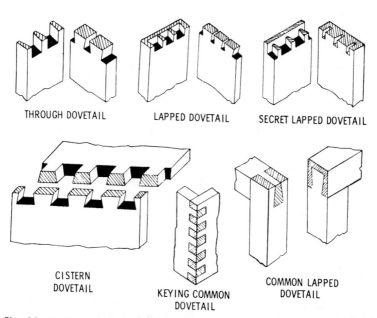

THROUGH DOVETAIL   LAPPED DOVETAIL   SECRET LAPPED DOVETAIL

CISTERN
DOVETAIL

KEYING COMMON
DOVETAIL

COMMON LAPPED
DOVETAIL

Fig. 33. Various types of dovetail joints. Dovetail joints are used principally in cabinetmaking, drawer fronts, and fine furniture work. They are a partly housed and tapered form of tenon joint in which the taper forms a lock to hold the parts securely together.

Common Dovetail Joint—This is a plain, or single "pin," joint. In dovetail joints, the tapered tenon is called the *pin*, and the mortised part that receives this joint is called the *socket*. Where strength rather than appearance is important, the common dovetail joint is used. The straight form of this joint is shown in Fig. 34, and the corner form is shown in Fig. 35; the proportions of the joint are shown in Fig. 35A and B.

Compound Dovetail Joint—This is the same as the common form but has more than one pin, thereby adapting the joint for use with wide boards. When making this joint, both edges are made true and square; a gauge line is run around one board at a

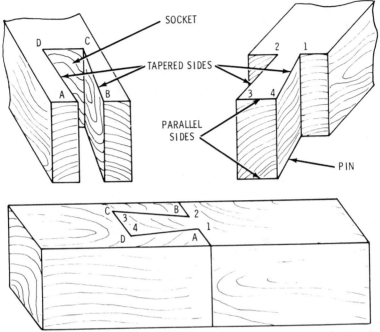

Fig. 34. The straight form of the common, or plain, dovetail joint. By noting the positions of the letters and numbers, it may be seen how the socket and pin are assembled.

distance from the end equal to the thickness of the other board, and the other board is treated similarly. Two methods are commonly followed. Some mark and cut the pins first; others mark and cut the sockets first.

In the first method, the pins are carefully spaced, and the angles of the tapered sides are marked with the bevel. Saw down to the gauge line, and work the spaces in between with a chisel and a mallet. Then, put B on top of A (in Fig. 36), and scribe the mortise. Square over, cut down to the gauge line, clean out, and fit together.

The second method is to first mark the socket on A (sometimes on common work, the marking is dispensed with, and the worker uses his eyes as a guide); then, run the saw down to the gauge line, put A on B, and mark the pins with the front tooth of

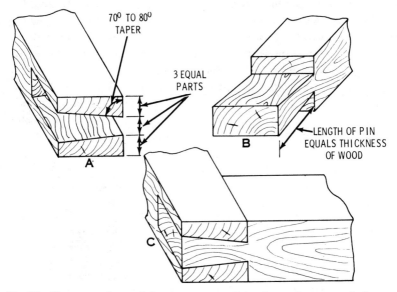

Fig. 35. The corner form of the common, or plain, dovetail joint, with the proper proportions for the socket and pin.

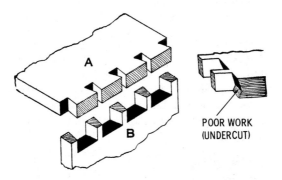

Fig. 36. A multiple dovetail joint, with a poorly cleaned joint shown in detail.

the saw. Cut the pins, keeping outside of the saw mark sufficiently to allow the pins to fit tightly; both pieces may then be cleaned out and tried together.

When cleaning out the mortises and the spaces between the pins, the woodworker must cut halfway through, then turn the

293

board over and finish from the other side, taking care to hold the chisel upright so as not to undercut, as shown in Fig. 36, which is sometimes done to ensure the joint fitting on the outside.

Lap or Half-Blind Dovetail Joint—This joint is used in the construction of drawers in the best grades of work. The joint is visible on one side but not on the other, as shown in Fig. 37, hence the name "half blind." Since this form of dovetail joint is used so extensively in the manufacture of furniture, machines have been devised for making the joint, thus saving time and labor.

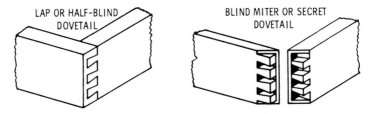

Fig. 37. Half-blind and blind dovetail joints, respectively. These joints were used in the best grades of drawer and cabinetwork, since the joint is visible on only one side. They had to be exceptionally well fitted because of the frequent pull on the front piece.

Blind Dovetail Joint—This is a double lap joint; that is, the joint is covered on both sides, as shown in Fig. 37, and is sometimes called a secret dovetail joint. The laps may be either square, as in Fig. 38, or mitered, as in Fig. 39. Because of the skill and time required to make these joints, they are used only on the finest work. The mitered form is the more difficult of the two to assemble.

Spacing—The maximum strength would be gained by having the pins and sockets equal; however, this is rarely done in practice, since the mortise is made so that the saw will just clear at the narrow side with the space from eight to ten times the width of the widest side. Small pins are used for the sake of appearance, but fairly large ones are preferable. The outside pin should be larger than the others and should not be too tight or there will be the danger of splitting, as shown in Fig. 40 at point A. The angle of taper should be slight (70° to 80°) and not acute as

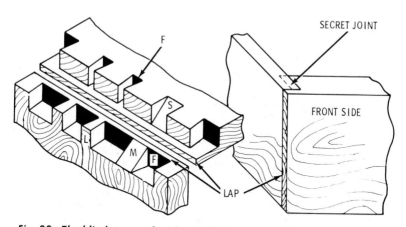

Fig. 38. The blind, square-lap dovetail joint was another useful joint. Two forms of pins and sockets were used—mitered (MS) and square (LF).

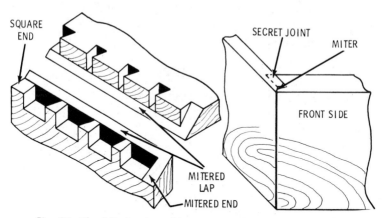

Fig. 39. The blind, mitered-lap dovetail joint, another oldie.

shown, otherwise there is the danger of pieces L and F in Fig. 40 being split off in assembling.

Position of Pins—When boxes are made, the pins are generally cut on the ends with the sockets on the sides. Drawers have the pins on the front and back. The general rule is to locate the tapered sides so that they are in opposition to the greatest stress

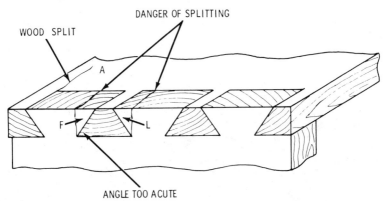

Fig. 40. A badly proportioned common dovetail joint can result in splitting.

that may be applied on the piece of work to which the joint is connected.

Tongue-and-Groove Joint—In this type of joint, the tongue is formed on the edge of one of the pieces to be joined, and the groove is formed in the other, as shown in Fig. 41.

SUMMARY

In carpentry and woodworking the term "joint" means the union of two or more smooth or even surfaces. The goal is to

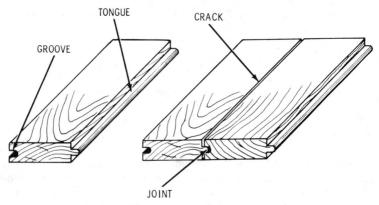

Fig. 41. The tongue-and-groove joint.

achieve a strong joint without weakening any part of the rest of the structure, by removing too much stock.

There are many wood joints, which may be divided into the classifications such as plain or butt joints, lap joints, mortise-and-tenon joints, and dovetail joints, to name a few. A plain or butt joint is where the end or one side of a piece is placed or butted against one end or side of the other. A lap joint is where two pieces to be joined lap over or into one another.

Plain or butt joints are generally classified as straight or plain-edge, dowel pin, splined or feather, beveled spline or miter, and beveled plain edge. Straight or plain-edge joints are the simplest form of joint and have many uses where several pieces are required to form a flat surface.

Dowel joints could be considered as a substitute for mortise-and-tenon joints. If well made and not exposed to weather and extreme temperature changes, it is a strong and excellent joint. A dowel joint is simply a butt joint reinforced by dowels which fit tightly into holes bored in each member to align them with each other.

There are many variations of the mortise-and-tenon joint. Among the more common joints are stub tenon, through tenon, haunched tenon, open tenon, and double tenon. When cutting a mortise, select a chisel as near the width of the mortise as possible. Check all cuts for accuracy before applying glue.

REVIEW QUESTIONS

1. What is a plain or butt joint?
2. What is a mortise-and-tenon joint?
3. What is a dovetail joint?
4. What are dowel joints? How do they compare with mortise-and-tenon joints?
5. Name the five mortise-and-tenon joints.

CHAPTER 22

Cabinetmaking Joints

In making cabinets and other fine furniture, the name of the game is joinery—good joints that hold the parts together well. This chapter explores the tools, materials, techniques, and joints that accomplish this goal.

THE TOOLS

A full set of good carpenters' tools is necessary in the cabinetmaking shop, including a set of firmer chisels, a set of iron bench planes, a set of auger bits with slow-feed screws that range in size from $1/4$ to 1 inch, and a set of numbered bits for the electric drill. In addition, the following tools will be found useful almost continually.

1. Router plane.
2. Plow plane.
3. Miter box.
4. Several bar clamps with varying lengths of bars.
5. Hand clamps (those with wood jaws are most useful in the shop, but malleable C clamps are often used).
6. A $1/4$-inch electric drill.
7. A chute board of wood or iron.

In addition to the regular vise, a workbench with an end vise for holding material between stops on top of the bench will be found quite convenient.

JOINTS

The great variety of joints used in cabinetwork are usually classified according to their general characteristics, as glued, halved and bridle, mortise-and-tenon, dovetail, and mitered. Under each of these classifications is grouped a variety of joints that will be considered separately and will also be briefly explained.

Glued Joints

In cabinetwork, practically all joints are, or should be, glued. There are several types of glues suitable for use in the cabinet shop.

White Glue, or Polyvinyl—leads the list. It is a most common woodworking glue. It is easy to apply, dries in about an hour and is clear. Like most glues, it must be clamped until set.

Carpenter's Glue—a relative newcomer on the market. It works very much like white glue, but it dries much faster, in perhaps half the time.

Contact Cement—the adhesive favored for the application of plastic laminate. The adhesive is applied to both surfaces or things to be joined, allowed to get tacky, then pressed together. As the name suggests, the parts bond together instantly— indeed, there is really no room for misalignment errors.

Casein Glue—This is a glue with a lot of gap-filling properties, and it works well when temperatures are low. It also works well in bonding oily woods such as teak, which could create problems for other glues.

Hot-Melt Glue—This is another relative newcomer on the market. It is applied with a "gun." Solid stocks of glue are inserted in a chamber in the gun where they are heated to a hot-metal state. The glue can then be extruded on the wood. The great advantage of hot-melt glue is that no clamping is required. Just push the items together, and the bond occurs within 30 to 60 seconds, depending on the brand of gun you buy.

Epoxy—This is really not a woodworker's glue, but it is very strong and you should know about it. Epoxy comes in two parts that are mixed together before use. It can be used to bond china, glass, and a variety of other materials, including wood.

Whatever glue you use, the parts to be mated should be clean, and the glue should be applied as recommended by the manufacturer. Good pressure must be applied, and the mating parts must be in solid contact. If glue oozes out of a joint, wipe it away immediately.

Beveled Joints

In this type of joint, the sides of the pieces fit together to form angles, or corners, as shown in Fig. 1. An infinite amount of planing and dressing can be saved by first ripping the edges roughly. This can be done by hand, of course, but you may want to take advantage of a table saw. If ripped on a power saw, the angle can be adjusted precisely, and only a small amount of hand dressing will be necessary. Try the bevel continually with a T-bevel while dressing to assure that the joints fit; they must fit properly if the joint is to be glued. The joints must be clamped, and without special clamps this is troublesome. The woodworker's ingenuity will usually suggest a method. Short pieces of chain with bolts through the end links are useful, if the chains can be passed around the work. Beveled joints do not usually require exceedingly high pressure.

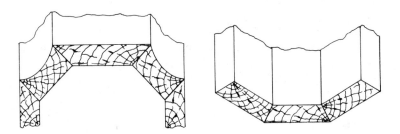

Fig. 1. Typical beveled joints.

Plowed-and-Splined Joints

This method of jointing is used in cabinetwork and is similar to the spline joint in the preceding chapter, except that in cabinetwork, the splines are cut *across* the grain. When the thickness of the material will permit, two splines are used instead of one, as in Fig. 2B, because of the additional gluing sur-

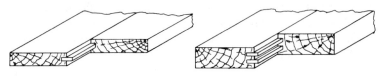

A. Single spline.   B. Double spline.
Fig. 2. Single- and double-splined joints.

face afforded and the increased strength to the joint. Splines are cut lengthwise with the grain.

To make this joint successfully, the pieces should be properly faced, and the edges should be squared and straightened with a jointer so that they fit perfectly. Put reference marks on the face side so that the same edges will come together when assembled. Set the plow plane with the iron projecting approximately $^1/_{32}$ inch below the bottom plate. Set the depth gauge to one-half the width of the spline, and adjust the fence so that the cutter will be the required distance from the edge between the two sides. Fasten the piece securely in the bench vise so that the groove can be plowed from the face side. Begin plowing at the front (Fig. 3A) and work backward; finish by going right through from back to front, as in Fig. 3B. Hold the plow plane steady, otherwise an irregular groove will result.

For the cross-grain spline, cut off the end of a thin board of hardwood; mark it, and carefully saw off a strip across its width that is the required width of the spline, approximately ¾ inches wide. Plane the spline to the desired thickness in a tonguing board. Then, assemble the parts and glue them up, as shown in Fig. 4.

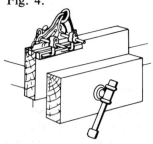 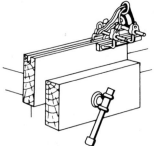

A. Starting.   B. Finishing.
Fig. 3. The method of plowing a single spline groove.

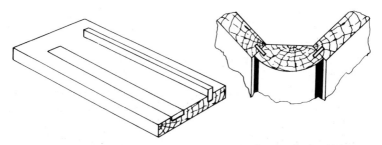

A. A tonguing board.   B. Plowed-and-splined joints.

Fig. 4. The tonguing board is a simple and handy device when used to overcome the difficulty of holding a narrow piece of thin material steady while planing. To make the board, use a piece of ⅞-inch faced material, 8 to 10 inches wide and longer than the tongues to be planed. Cut the grooves, as indicated, with a tenon saw; clean out the grooves with a chisel and a router. The wider groove should be slightly deeper than the thickness of the finished tongue to allow for planing both sides of boards placed in it.

Hidden Slot Screwed Joints

This joint is not often used as a glued joint but is found to be an effective way of fastening brackets and shelves to finished work where the fastening must be concealed. The joint consists of a screw which is driven part way into one piece and a hole and slot cut into the opposite piece. The joint is effected by fitting them together with the head of the screw in the hole and then forcing the screw back into the slot. Fig. 5A shows the slot

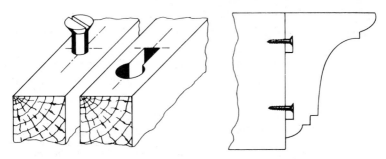

A. The joint before assembly.   B. Cross-sectional view of assembled
                                            joint.

Fig. 5. The hidden slot screwed joint.

and screw in relation to each other, and Fig. 5B gives a cross-sectional view of the completed joint. This joint is also used in interior work for fastening pilasters and fireplaces to walls, for panelling, and for almost every kind of work requiring secure and concealed fastening.

To make a joint as shown in Fig. 5B, gauge a center line on each of the pieces. Determine the position of the screws and insert them; they should project approximately ³/₈ inch above the surface. Hold the two pieces evenly together, and, with a try square, draw a line from the back of the screw shank across the center line of the opposite piece. From this line, measure ⁷/₈ inch forward on the center line; with this point as the center, bore a hole to fit the screw head that is just slightly deeper than the amount that the screw projects above the surface. Cut a slot from the hole back to the line from the screw shank; this slot should be as wide as the diameter of the shank and as deep as the hole. As a general rule, the total length of the slot and hole should be slightly more than twice the diameter of the head of the screw.

The process of fastening pilasters to fireplaces by this method is as follows: First, mark the position of the piece and the place on the wall for the screws. In brick and cement walls, holes are drilled and wooden plugs are driven in flush with the surface to hold the screws. The plugs are shaped as illustrated in Fig. 6. Plugs cut as shown seldom work loose. Turn the screws into the plugs, and allow them to project approximately ³/₈ inch from the surface; the screw heads should be smeared with a bit of lampblack. Put the piece in position and press it against the screw heads; this pressure will leave black impressions. Bore holes to fit the screw heads approximately ⁵/₈ inch below the impressions, and cut the slot to receive the shank of the screw.

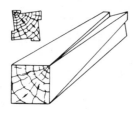

Fig. 6. A plug is used to hold the screw when hidden slot screwed joints are utilized on mortar and brick walls.

Replace the piece, with the heads in the holes, and force it down.

Dowel Joints

There are many variations of dowel jointing commonly used in cabinetwork. The basic principle and method of making a dowel joint is explained and illustrated in the preceding chapter.

To accurately fix the position of dowels in a butt joint, such as the one shown in Fig. 7, make all measurements and gauge lines from the edge of the faced sides. For example, with material that is 4 inches square, mark diagonal lines from the corners with a scratch awl, intersecting at the center. Then, from the edge of the faced sides, mark off 1 inch and 3 inches as shown in Fig. 8A. From the same sides, gauge the lines as in Fig. 8B. The intersection of the lines is the center for the holes, as shown in Fig. 8C.

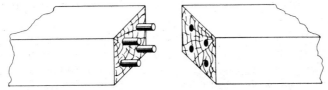

Fig. 7. The use of dowels in a butt joint adds strength to the joint. This type of construction is frequently found in cabinet work to lengthen large mouldings and where the cross grain prevents tenoning.

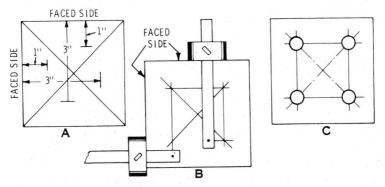

Fig. 8. The method used for marking the position of the dowels in a butt joint.

Where the ordinary means of aligning dowel holes cannot be used, a dowel template or pattern is used. The template is usually made of a strip of zinc or plywood, with a small block of wood fastened to one end to act as a shoulder; the position of the dowel points is then pierced through the zinc or plywood pattern with a fine awl. Various types of templates are made and used as the occasion requires. Fig. 9A shows a template that is used for making dowel rails in furniture; it is made to fit the section of the rail in Fig. 9B. While held in position, a line is gauged down through the middle; the position of the dowels is indicated on the line. The template is laid flat on a board, and the dowel points are pierced through the surface with a fine awl. When in use, the template is placed in position on the piece, and the dowel positions are marked with an awl through the holes, as shown in Fig. 9C. A "bit gauge" should be used to regulate the depth of the bore when doweling. If a great amount of doweling is to be done, a doweling jig, which ensures accurate boring of holes from $\frac{1}{4}$ to $\frac{3}{4}$ inch, will be found useful.

Dowels are glued into one piece, cut to length, and sharpened with a dowel sharpener. As a precaution against splitting the joint, cut a V-shaped groove down the side of the dowel with a chisel; this groove permits the glue and air to escape.

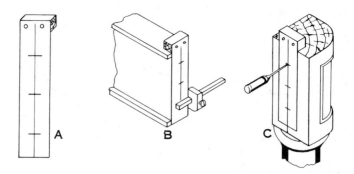

Fig. 9. The use of a template, or pattern, for marking dowel-pin locations; A, the template; B, the template is made to fit the section of rail; C, marking dowel positions on leg.

Coopered Joints

These are so named because of the resemblance to the joints used in barrels made by coopers, and are used for practically all forms of curved work; they are usually splined before gluing, although dowels are used occasionally. Fig. 10 shows the coopered joint in semicircular form, with the segments beveled at an angle of 15 degrees. They are clamped after gluing and planed to shape.

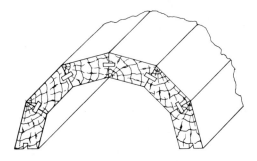

Fig. 10. Coopered joints are employed to form various curvatures in cabinet work.

Halved and Bridle Joints

These are lap joints with each of the pieces halved and shouldered on opposite sides so that they fit into each other. They are the simplest joint used in cabinetwork. Fig. 11A shows the common halved angle, which is the one most frequently used. Fig. 11C illustrates the oblique halved joint, which is used for oblique connections. Fig. 11D represents the mitered halved joint, which is useful when the face or frame piece is moulded. Fig. 11E, F, and G shows the joints that are used for cross connections having an outside strain. Fig. 11H illustrates the blind dovetail halved joint, which is used in places where the frame edge is exposed. Bridle or open-tenon joints are used to connect parts of flat and moulded frames. The joint in Fig. 12B is used where a strong framed groundwork, which is to be faced up, is required. The joint in Fig. 12D is used as an inside frame connection.

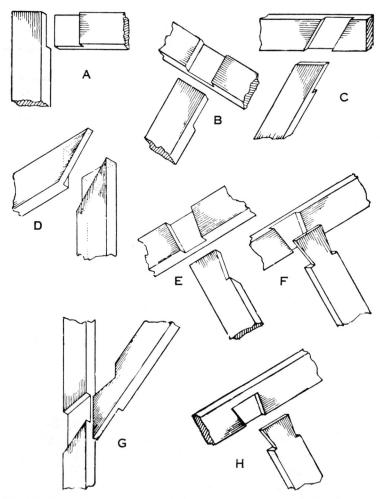

Fig. 11. Various halved joints; A, halved angle joint; B, halved tee joint; C, oblique halved joint; D, mitered halved joint; E, dovetail halved joint; F, dovetail halved joint; G, oblique dovetail halved joint; H, blind dovetail halved joint.

Mortise-and-Tenon Joints

Many variations of the mortise-and-tenon joints are used in cabinetwork. They differ in size and shape according to the requirements of the location and the purpose for which the joint is

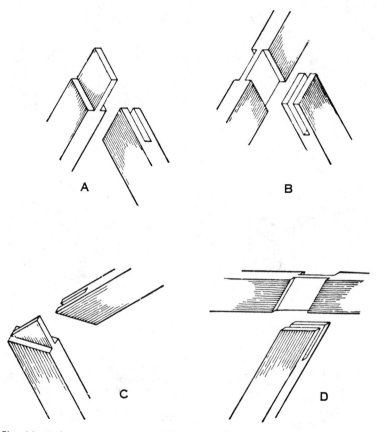

Fig. 12. Halved joints that are generally used to make flat and molded frames; A, angle bridle joint; B, tee bridle joint; C, mitered bridle joint; D, oblique bridle joint.

used. The most frequently used mortise-and-tenon joint is the stub tenon, so called because it is short and penetrates only part way through the wood; Fig. 13A illustrates the type most generally used in doors and furniture framing. Fig. 13B shows the stub tenon with a mitered end; this type of construction is often necessary when fitting rails into a corner post. Fig. 13C shows the rabbeted or ''haunched'' tenon, which is considered a stronger joint because of the small additional tenon formed by the rabbet; it is often mitered, as shown in Fig. 13D, to conceal

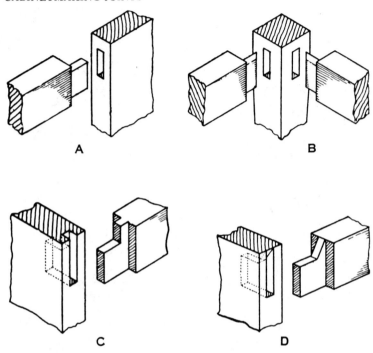

Fig. 13. Variations of the mortise-and-tenon joints are frequently used in doors and in the framing of furniture; A, stub mortise and tenon; B, mitered stub tenon; C, haunched, or rabbeted, mortise and tenon; D, haunched and mitered mortise and tenon.

the joint when used on outside frames. The joint in Fig. 14A is the same as that shown in Fig. 13C, but it is shouldered on one side only; it is sometimes called a "barefaced" tenon and is used when the connecting rail is thinner than the stile into which it is joined. Fig. 14B shows the long and short shoulder tenon; this joint is used when connecting a rail into a rabbeted frame, since it has one shoulder cut back so as to fit into the rabbet. Fig. 14C illustrates the double tenon joint, which increases the lateral strength of the stile into which it is jointed. It is simply a stub tenon that is rabbeted and notched to form two tenons, and, when glued, it makes an exceptionally strong joint. Fig. 14D represents a type of through mortise and tenon that is sometimes used for mortising partitions into the top or bottom of ward-

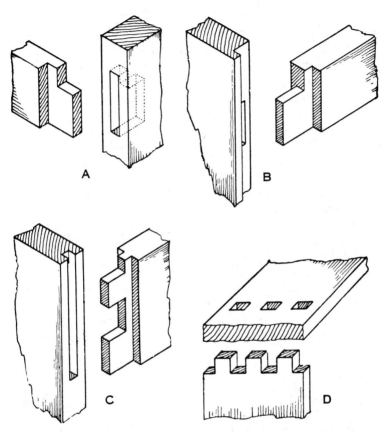

Fig. 14. Other mortise-and-tenon joints used in the construction of furniture; A, barefaced mortise and tenon; B, long and short shoulder mortise and tenon; C, double mortise and tenon; D, pinned mortise and tenon.

robes, cabinets, etc.; the partitions are wedged across the tenon and glued.

Laying Out the Mortise and Tenon—The general practice when laying out a mortise and tenon is to square the mortise lines across the edge of the stile in pencil and then scribe two lines for the sides of the mortise with a mortise or slide gauge between the pencil lines. If the tenon is to be less than the full width of the rail, square the rail lines across the edge, in addition

311

to the mortise lines, as shown in Fig. 15A. This procedure ensures greater accuracy when designating the position of the mortise. When two or more stiles are to be mortised, they are clamped together, and the lines are squared across all the edges simultaneously.

For a through mortise, continue the pencil lines across the face side and onto the back edge; gauge the mortise lines from the faced side. With the gauge set for the mortise, scribe the lines for the tenon on both edges and the end of the rail, and with the aid of a try square, mark the shoulder lines with a knife or chisel on all four sides.

The proportions of stub and through mortises and tenons are usually considered as about $\frac{1}{3}$ the thickness of the wood, and they should be cut with a sharp chisel of the required size. If the chisel is not exactly $\frac{1}{3}$ the thickness of the material, it is better to make the mortise more than $\frac{1}{3}$ rather than less. Set the mortise gauge so that the chisel fits exactly between the points, as shown in Fig. 15B. Make a chisel mark in the center of the edge to be mortised, and adjust the head of the gauge so that the points coincide with this mark.

Mortise cutting in cabinetwork is usually done entirely with a sharp chisel, beginning at the center and working toward the

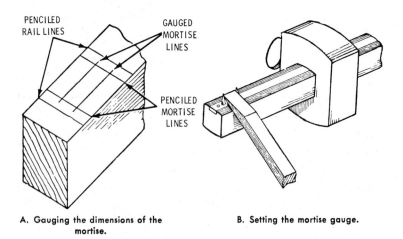

A. Gauging the dimensions of the mortise.

B. Setting the mortise gauge.

Fig. 15. Laying out the mortise.

near end with the flat side of the chisel toward the end. Remove the core as you proceed; then reverse the chisel, and cut to the far end, being careful to keep the chisel in a perpendicular position when cutting the ends. Through mortises are cut halfway through from one side, and the material is then removed and cut through from the opposite side.

A depth gauge for stub mortises is made by gluing a piece of paper or tape on the side of the chisel, as shown in Fig. 16. If the method of boring a hole in the center from which to begin the cutting of the mortise is used for stub mortises, it is advisable to use a bit gauge to regulate the depth of the bore. A small firmer chisel is used to clean out stub mortises.

Fig. 16. The depth of the mortise joint may be controlled by fastening a piece of paper or tape on the side of the chisel.

PAPER
DEPTH GAUGE

Cutting the Tenon—Fasten the piece firmly in the bench vise. Start the cut on the end grain, and saw diagonally toward the shoulder line, as in Fig. 17A. Finish by removing the material in the vise and cutting downward flush with the edge, as in Fig. 17B. The diagonal saw cut acts as a guide for the finishing cut and provides greater accuracy. Small tenons are usually cut with a dovetail saw.

Cutting the Shoulder—After making the tenon cuts, and to overcome any difficulty in cutting the shoulders, place the piece on the shoulder board or bench hook, and carefully chisel a V-shaped cut against the shoulder line, as shown in Fig. 18A. Hold the work firmly against the stop on the board; place the saw in the chiseled channel, and begin cutting by drawing the saw backward and then pushing it forward with a light stroke. Hold the thumb and forefinger against the saw, as in Fig. 18B,

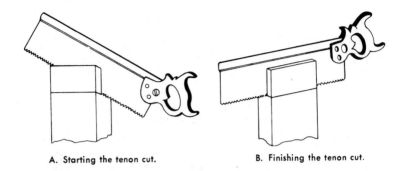

A. Starting the tenon cut.

B. Finishing the tenon cut.

Fig. 17. Making a tenon cut.

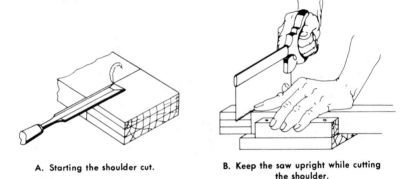

A. Starting the shoulder cut.

B. Keep the saw upright while cutting
the shoulder.

Fig. 18. Making the shoulder cut.

and keep the saw in an upright position. A straightedge can be placed against the shoulder line to act as a guide when cutting wide shoulders. In the case of extremely wide tenons and shoulders, a rabbet plane and a shoulder plane are used; the straightedge is used as a guide for the rabbet plane.

Dovetail Joints

The method of making dovetail joints is described in the previous chapter. In common dovetailing, it is a matter of convenience whether to cut the pins or the dovetails first. However, where a number of pieces are to be dovetailed, time can be saved by clamping them together in the vise and cutting the dovetails first.

Dovetail Angles—For particular work where the joint is exposed, the dovetails should be cut at an angle of 1 in 8, and for heavier work, 1 in 6. To find the dovetail angle, draw a line square with the edge of a board, and divide it into 6 or 8 equal parts as desired; from the end of the line and square with it, mark off a space equal to one of the divisions, and set the bevel as shown in Fig. 19.

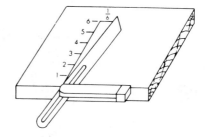

Fig. 19. Finding the angle of the dovetail.

A dovetail template, as shown in Fig. 20A, will be found quite handy if there is a great deal of dovetailing to be done. To make the template, take a rectangular piece of ³/₄-inch material of any desired size, and square the edges; with the mortise gauge set for a ¹/₄-inch mortise at ¹/₄ inch from the edge, scribe both edges and one end. With the bevel set as shown (Fig. 19), mark the shoulder lines across both sides of the lower portion, and cut it with a tenon saw. Make one cut for each of the two angles. The template may also be made by gluing a straightedge, at the required angle, across both sides at one end of a straight piece of thin material. The use of a dovetail template saves time and ensures uniformity. Place the shoulder of the template against the edge, as shown in Fig. 20B, and mark one side of the dovetail along its edge. Reverse the template, place the other shoulder at the same edge, and mark the other side of the dovetail.

Beveled Dovetailing—The joint shown in Fig. 21 is sometimes required in cabinetwork, and a template can be a great help for marking it. To use the template for marking beveled dovetails, cut a wedge-shaped piece of material, as shown in Fig. 22A, that

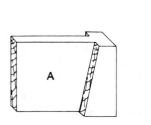

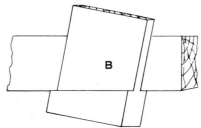

A. The dovetail template.

B. The dovetail template in use for layout work.

Fig. 20. A template is invaluable for dovetail work.

Fig. 21. The beveled dovetail joint.

is beveled at the same angle as the bevel of the material to be dovetailed. Insert this wedge between the edge of the material and the template, with the square edge of the wedge against the shoulder of the template, as in Fig. 22B. Mark the dovetail as described, but do not reverse the wedge-shaped piece.

A

B

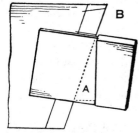

A. A wedge is used with the dovetail template to mark the desired bevel.

B. Laying out beveled dovetails with the template and wedge.

Fig. 22. The method of laying out work with the aid of a dovetail template.

The common, or through dovetail, shown in Fig. 23A, is primarily used for dovetailing brackets and frames which are subject to a heavy downward strain. Fig. 23B illustrates the common lapped, or half-blind, dovetail as it is applied to a curved door frame; it is used in all locations of this type where mortise-and-tenon joints would not be effective. The common lapped dovetail joint may also be used for purposes similar to those described for the common dovetail joint.

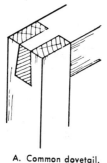

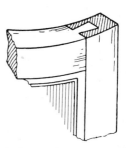

A. Common dovetail.

B. Common lapped dovetail.

Fig. 23. Two typical dovetail joints.

Fig. 24A illustrates the common housed "bareface" dovetail; it is shouldered on one side only. The joint in Fig. 24B is shouldered and dovetailed on both sides and is another of the same type with the dovetailing parallel along its entire length. These are the simplest forms of housed dovetailing. Their application to the framing of furniture is shown in Fig. 29, under the section on framing joints discussed later in this chapter.

Fig. 24C illustrates a shouldered housing dovetail joint with the dovetail tapering along its length; as with the two preceding joints, this joint can be shouldered on one side or both sides. The tapered dovetail makes this joint particularly adaptable for connecting fixed shelves to partitions, because the dovetails prevent the partitions from bending. A dovetailed and housed joint, frequently called a "diminished" dovetail, is shown in Fig. 24D; it is principally used on comparatively small work, such as small fixed shelves and drawer rails.

Making a Diminished Dovetail—Square division lines across

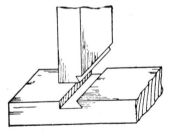

A. Barefaced dovetail housing.

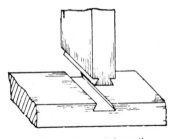

B. Common housed dovetail.

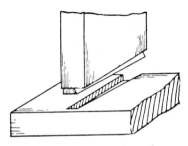

C. Shouldered housing dovetail.

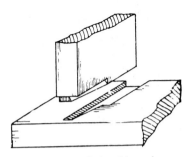

D. Dovetailed and housed.

Fig. 24. Housed dovetails of the single- and double-shouldered types.

the ends into which the shelf is to be housed and dovetailed as far apart as the thickness of the shelf, and gauge the depth of the housing on the back edge. Gauge lines $3/8$ and $4^1/_2$ inches from the front edge between the division lines; the space between these gauge lines is the length of the actual dovetail, as shown in Fig. 25A.

Cut out the section indicated at **A** with a chisel, and undercut side **B** to form a dovetail; insert a tenon saw, and cut the sides across to the edge. Remove the core with a firmer chisel, and finish to depth with a router. Gauge lines on both ends of the shelf on the side and end for the depth of the dovetail, and square across the end the distances from the front edge as given; cut away the surplus wood with a tenon saw, as shown in Fig. 25B, and finish the cut with a chisel, carefully testing until it fits hand tight. Fig. 25C shows the completed end. The average length of the actual dovetail of this type is slightly less than $1/4$ of the total length of dovetail and housing.

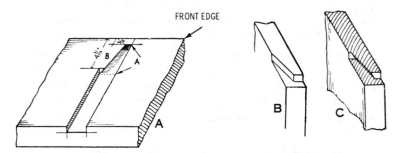

Fig. 25. The method of making a diminished dovetail joint.

Mitered Joints

Mitering is an important part of cabinetwork, in the framing of furniture, and in panelling, where many difficult mouldings must be mitered into place. Fig. 26A illustrates a plain miter with a cross tongue (or spline) inserted at right angles to the miter. This joint is principally used for mitering end grain and is additionally strengthened by gluing a block to the internal angle, as shown in Fig. 29.

Tonguing a Miter—One practical way of tonguing this joint is to fasten two miters together in a vise so as to form a right angle, as in Fig. 26B, thus providing an edge from which to gauge the position of the tongues and plow the grooves. If the pieces are not over 6 inches in width, the grooves are cut with a dovetail saw and chiselled to depth.

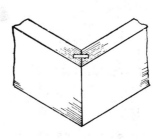

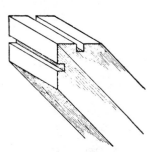

A. A tongued miter.   B. Method of tonguing a miter.

Fig. 26. Mitered joints.

319

Fig. 27A is a variation of plain mitering and, like the preceding joint, is most generally used for end-grain jointing. For this joint, the tongue should be approximately $\frac{1}{3}$ the thickness of the material, and it may extend all the way through or only part way, as shown in Fig. 27B. This joint is especially useful in cabinetwork for connecting and mitering various types of large mouldings around the tops of pieces, for mitering material for tops and panels, and for connecting sections in curved work, as shown in Fig. 27C.

Screwed Miter Joint—Fig. 28A illustrates a plain miter with a screw driven at right angles to the miter across the joint through

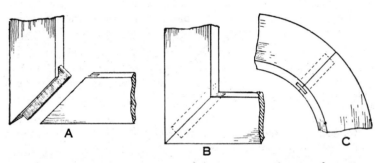

Fig. 27. The tongued miter joint in various stages and types of construction; A, before completion; B, completed joint; C, for connecting segments in curved work.

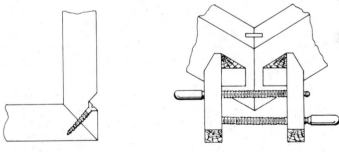

A. Cross-sectional view of the completed joint.

B. Method of clamping a miter joint.

Fig. 28. The screwed miter joint.

a notch cut in the outside of the frame. This type of joint is used principally in light moulded frames.

A common method of clamping a tongued miter is to glue blocks to the piece and hand-screw the joint together, as shown in Fig. 28B. The blocks are glued on and allowed to dry before gluing up the joint; when the joint is dry, the blocks are knocked off, and their marks are erased.

Fig. 29 is a part plan for the base of a breakfront cabinet; it shows the application of mitered and housed dovetail joints to furniture construction.

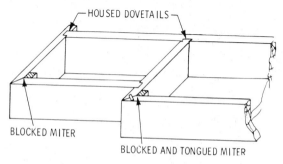

Fig. 29. The application of various forms of joints in cabinet construction.

Framing Joints

The term "framed," or "framing," as used in cabinetwork, indicates work that is framed together, as in Fig. 29. It also refers to the "grounds" for securing panelling to walls.

Fig. 30 represents various joints used to connect angles for panelling. The joints shown in Fig. 30A and B are identical except for the return bead, which is worked on one of the pieces. These two joints are usually glued and nailed, but, when used as an external angle, they are countersunk before painting; that is, the screws are sunk below the surface, and plugs or pellets of wood are glued in the holes and beveled off. The joint represented in Fig. 30C may be used to connect framing at any angle; the rabbet prevents slipping while being nailed or screwed. Fig. 30D shows an ordinary rabbeted joint with the corner rounded off and the pieces glued together; because of its rounded corner, it is often used in furniture for children's nurseries. Fig. 30E

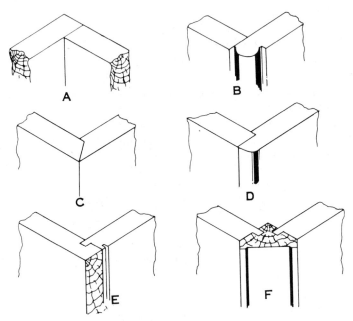

Fig. 30. Commonly used framing joints; A, butt, or square joint; B, return bead and butt joint; C, rabbet and miter joint; D, rabbet and round joint; E, barefaced tongued joint; F, splayed corner joint.

illustrates a joint that is shouldered on one side only. A bead is worked on the tongue piece to hide the joint. It is used for both internal and external angles, with or without the bead. The joint in Fig. 30F is the splayed corner tongued joint, which is used for joining sides into a pilaster corner.

SUMMARY

A great variety of joints used in cabinetwork are usually classified according to their general characteristics, such as glued, halved and bridle, mortise and tenon, dovetail, miter, framing, and hinging and shutting. In cabinetwork, all joints should be glued. All glued joints must be clamped, or pressure applied by means of nails or screws.

The glue that is the most popular in woodworking shops is polyvinyl. This glue does not stain wood; it is light in color, easy to use, and can set up at room temperature. Good glued joints must be in close contact while they are setting up. If this contact is not made, the joint will not have the strength required to hold up under normal service.

REVIEW QUESTIONS

1. Name a few of the tools used in cabinetmaking joints.
2. Why is it so important to clamp glued joints?
3. What type of glue is best for general glued joints?
4. What is a coopered joint?
5. What is a dovetail joint?

CHAPTER 23

Wood Patternmaking

The term "patternmaking," as used here, means the making of patterns to be used in forming molds for castings. Patterns are made of different materials, such as metal, plaster, and wood, with most made of wood. Indeed, patterns for every known kind of machine, from the smallest to the largest, are and can be made of this material.

Patternmaking is a highly skilled branch of carpentry requiring, as it does, proficiency in all aspects of the trade—from joining to carving to turning—the ability to read complicated blueprints, and a solid understanding of foundry and core work. While the average carpenter or handyman doesn't require a knowledge of patternmaking, it is something a person may desire to master. For this reason, the techniques of patternmaking are described here.

Although a pattern is defined as a model, its outward appearance does not always closely resemble the casting itself, except in a simple casting where the pattern is often a complete model. If, however, the casting is to have interior passages and external openings, its appearance will be changed by the addition of projections called "core prints." These are placed so as to form bearing surfaces in the mold to support the sand cores used to form those passages or openings in the castings. These cores are usually formed in wooden molds called "core boxes," and the prints on the pattern are commonly distinguished by being painted a different color than the pattern itself. A pattern is given a certain amount of taper, or "draft," to ensure its easy removal from the molding sand; the removing process is called "drawing." A pattern is said to draw well or not draw well according to the amount of trouble experienced in removing it from the molding sand.

PATTERNMAKER'S TOOLS

The first requirement of a patternmaker is a complete set of good tools. The following list is adequate for a wide range of work:

1. Jack plane.
2. Block plane.
3. End-wood plane (for planing long end-wood edges; it is 14 inches long, with the iron ground straight along the cutting edge, often referred to as a jack plane).
4. Jointer plane.
5. Rabbet plane.
6. Router.
7. Circular plane (adjustable to both concave and convex surfaces).
8. Core-box plane.
9. Paring chisel (Fig. 1A).
10. Paring gouge (Fig. 1B).
11. Outside ground gouge (so called because it is ground on the outside, or convex, side; it is shorter and heavier than the paring gouge and is used for roughing work and with the mallet when necessary).
12. Carving tools (those most commonly used are known as straight, short-bend, and long-bend tools; a number of different sweeps of each style).
13. Bit brace.
14. Hand drill.
15. Electric drill.
16. Auger bits (in sizes from $3/16$ to 1 inch).
17. Gimlet bit (for drilling screw holes).
18. Center bit (for drilling thin stock).
19. Forstner bit (for drilling flat-bottom holes).
20. Expansion bit.
21. Bit stock drills (standard twist drills with square shank to fit hand braces, $1/16''$ — $1/4''$).

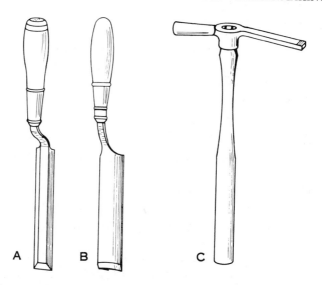

Fig. 1. The paring chisel (A), the paring gouge (B), and the pattern-maker's hammer (C) are three of the many tools used in patternmaking. The chisel varies in width from ⅛ to 2 inches, with square or beveled edges and straight or offset tangs; for general use, the ½-inch width is the most convenient, preferably with beveled edges and an offset tang, as illustrated. The paring gouge, like the chisel, is long and thin and varies in width from ⅛ to 2 inches, having three different curve sweeps known as "flat," "middle," and "regular." They are called inside-ground gouges, because they are ground on the inner, or concaved, side. The patternmaker's hammer is especially designed to meet the requirements of the trade; the long slender end is used for driving nails in fillets and for reaching corners that cannot be reached with the ordinary hammer.

22. Screwdriver bit.
23. Countersink bits.
24. Bit stop.
25. Backsaw.
26. Keyhole saw.
27. Coping saw.
28. Hammer (patternmaker's) (Fig. 1C).
29. Claw hammer (medium size).
30. Oilstones (and slip stones of different shapes for sharpening the edge of cutting tools).
31. Spokeshave (large and small).
32. Hand screws.

33. Pinch-dogs (Fig. 2).
34. Bar clamps.
35. Distance marking gauge (monkey gauge).
36. Panel gauge.
37. Turning tools.
38. Hermaphrodite calipers.
39. Outside calipers.
40. Inside calipers.
41. Dividers (large and small).
42. Trammels (with inside and outside caliper points).
43. Bevel square.
44. Combination square.
45. Screwdrivers.
46. Scribers.
47. Shrink rules.
48. For a full-fledged patternmaker's shop, power tools, both stationary and portable, are required simply because they allow you to produce patterns with precision and speed. Following is a variety of portable and stationary tools.
    a. Circular saw.
    b. Band saw.
    c. Jointer.
    d. Surfacer.
    e. Trimmer.
    f. Jigsaw.
    g. Sanding machines.
    h. Disc sander.
    i. Belt sander.
    j. Wood milling machine.
    k. Grinder.
    l. Lathe.

TRADE TERMS

There are certain terms in common use in the patternmaking shop and in the foundry that the prospective patternmaker should know. The more important of these are listed below and illustrated in Fig. 3.

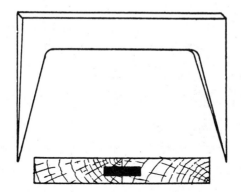

Fig. 2. A typical pinch-dog. The taper on the inside of the legs causes them to "pinch" when driven into two adjoining pieces.

Flask—A wood or metal frame made in two parts.

Cope—The top part of the flask.

Drag—The bottom part of the flask.

Bottom, or Mold, Board—A board which lies under the drag on which the pattern rests while the mold is being rammed. It is also used on the top while rolling the mold over.

Draft—The taper put on the pattern so that it will "draw" easily out of the sand.

Boss—A circular projection, knob, or stud.

Fillet—A round corner used in patterns.

MATERIALS

Lumber

The most common varieties of wood used in patternmaking are white pine, mahogany, cherry, maple, and birch. White pine is considered by far the best wood for all but the smallest patterns. It is soft and therefore easy to work; porous enough to take glue well, thereby ensuring strong glue joints; and, when properly seasoned, it is not greatly affected by exposure to heat, cold, or dampness. Cherry is used when there is great strength required in the pattern. It is also the wood most generally used for small patterns. Mahogany is used for patterns of light, thin

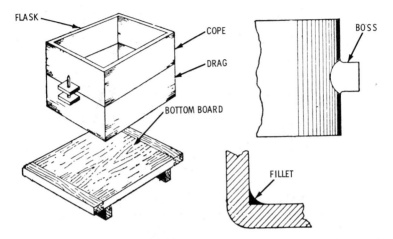

Fig. 3. The patternmaker's flask, with its components, and the boss and fillet, all of which are used in the patternmaker's trade.

construction that require a great amount of strength and hand working. Maple and birch are too hard for economical hand tool work, but they are well suited for small turned patterns. When a pattern is so large that it needs to be braced in the middle, a cheaper grade of wood may be used for the bracing. Pattern lumber should be carefully selected and should be free from knots and shakes (small cracks). The lumber must be thoroughly seasoned to prevent warping of the finished pattern.

Glue

Glue plays an important part in patternmaking; it is used for joining the different parts in patternwork. There are many different types of glue, each used for a specific purpose. The basic kinds of glue are:

1. Polyvinyl.
2. Resorcinol.
3. Epoxy.
4. Contact cement.

The surfaces of a joint should be made perfectly true before applying the glue. The surfaces that make contact should be dry, clean, and smooth prior to the spreading of the glue and the surface should be thoroughly and evenly coated. Although the application should be thorough, it should not be heavy. A heavy coat doesn't make parts adhere better and wastes material.

Shellac or Pattern Varnish

Yellow shellac (commercially known as orange shellac) is used to finish patterns to protect them from atmospheric moisture and the wet molding sand, which would warp them, and to make the patterns draw easier from the sand.

Pattern Colors

It is a general shop practice to indicate core prints and core-box faces by some given color and to paint the patterns with some recognized code of colors to correspond to the metals in which they are to be cast. For example:

Pattern and core-box bodies for iron casting are painted black.

Patterns and core-box bodies for steel casting are painted blue.

Patterns and core-box bodies for brass casting are painted orange.

All core prints and core-box faces are painted red.

Colored shellac is used for this purpose; it is made by dissolving the powdered pigment in alcohol and mixing it with the orange shellac.

Dowel Pins

Some patterns must be put together in such a way that certain parts can be easily removed and replaced again as required; a split pattern is a typical example. Small dowel pins of wood or metal are used for this purpose to assure proper alignment. Wooden dowel pins are used if the pattern is to be employed only a few times, but where there is much wear expected on a pattern and also on large patterns, metal dowels and dowel plates are invariably used. Two styles of metal dowel plates commonly used are shown in Fig. 4. They are secured with screws to the two parts of the pattern.

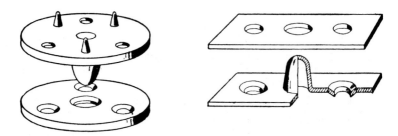

Fig. 4. Two types of metal dowel plates.

Fillets

Fillets, or concave connecting pieces, are used in the corners and at the intersection of surfaces of a pattern for the important reasons that they increase the strength of a casting by influencing the crystallization of the metal and also improve the appearance of the casting. Fillets are of two types—"stuck" and "planted." A stuck fillet is one that is cut out of the wood with a gouge, as shown in Fig. 5A. A planted fillet is one made separately and fitted in. Planted fillets are most commonly used and are made of wood, leather, and wax (beeswax).

Wood fillets are used for corners having large radii and on straight work. A round plane and a fillet board, Fig. 5B and C, are used for planing wood fillets. The stock is first reduced to a triangular section (Fig. 6A) on the circular saw and cut to suitable lengths; it is then put on the fillet board and planed with the round plane. A projecting screw head at the end of each groove acts as a stop. Wood fillets are always glued in place and are then securely nailed. To prevent the edges from curling, wet the face or concave side of the fillet with water before applying the glue. Thin strips are usually tacked over the shim, or feather, edges on large fillets, as in Fig. 6B, to hold them down until the glue is set, after which they are removed.

Leather fillets are strips of leather which are triangular in shape and are usually furnished in 4-foot lengths, as shown in Fig. 6C. They are extremely pliable and are easily attached to straight or curved work or to sharp or round corners, as shown in Fig. 6D. If moistened with water, their pliability is increased. They are fastened with glue or thick shellac and are rubbed in

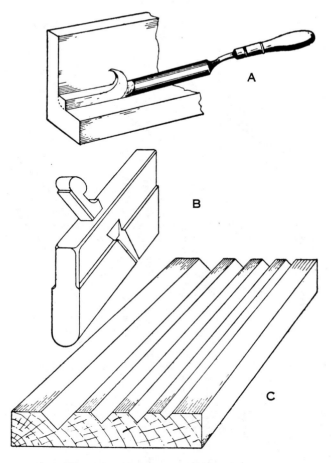

Fig. 5. Various fillet-shaping tools; A, cutting a fillet with a gouge; B, round plane; C, fillet board.

place with a waxing iron (Fig. 6E). If glued, the rubbing must be done rapidly and before the glue sets. The use of shellac permits more time for rubbing. Coat the back of the fillet as well as the corner to which it is to be applied; let the shellac get sticky before rubbing.

Beeswax fillets can be used on small patterns that are used infrequently but not if the pattern is likely to be molded in sand

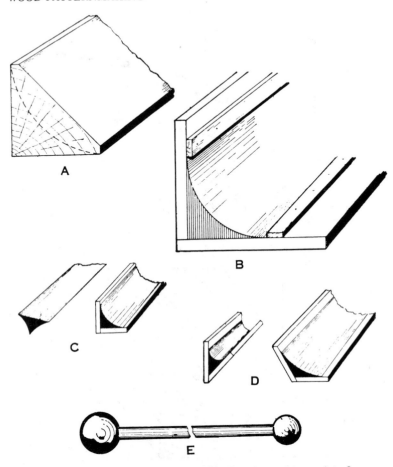

Fig. 6. Methods used to obtain different fillet shapes; A, stock is first reduced and then cut, as indicated by the dotted line, with a round plane;
B, application of wood fillets; C, leather fillets and their application; D,
leather fillets may readily be attached to sharp or rounded corners; E,
typical waxing iron used in the application of leather and beeswax fillets.

which has become warm enough to melt the wax. The fillets are
applied after the first coat of varnish. The wax is prepared for
use as fillets with a beeswax gun. The wax is forced out through
an opening in the side of the gun in the form of a long string that
is ready for use; it is rubbed into the corners with the waxing
iron.

TYPES OF PATTERNS

Generally speaking, patterns are divided into two classes—solid or one-piece patterns and split or parted patterns. In addition, there are some special types, such as skeleton patterns, part patterns, and patterns with loose pieces.

Solid or One-Piece Patterns

A solid or one-piece pattern is made without partings, joints, or any loose pieces; it is of one-piece construction when finished, and it may be a complete model of the required casting, or it may be partly cored, as shown in Figs. 7 and 8. Solid patterns require core prints for coring central holes when the hole is of such a size or shape that it cannot leave its own core in the molding sand, as, for example, the patterns shown in Fig. 9.

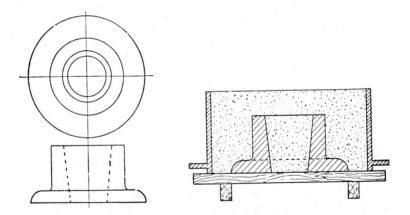

Fig. 7. A typical solid, or one-piece, pattern. It is a complete model of the required casting, including the control hole, which is large enough to let the molding sand stand in it as a green sand core to form the hole in the casting. The pattern is placed on the molding board with the larger side down and the drag part of the flask placed over it. The surface of the pattern is covered with facing or fine molding sand, which is rammed in and around it, and the drag is filled with coarser sand, which is rammed flush with the top. A bottom board is placed on top of the drag, which is then turned over; the first board is then removed. Parting sand is dusted over the exposed side of the drag; the cope side of the flask is put in place and is rammed flush with the top.

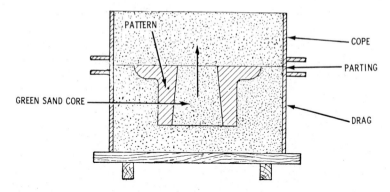

Fig. 8. The one-piece pattern and flask shown in Fig. 7. The vertical arrow indicates the direction in which the pattern is to be drawn from the mold.

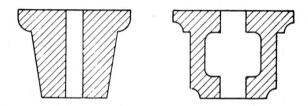

Fig. 9. Typical one-piece patterns.

Split or Parted Patterns

This is one that, because of its shape, cannot be drawn from the sand mold unless it is made in at least two parts, such as an engine cylinder with its flanges, shown in Fig. 10. Such a pattern will lie half in the drag, or lower part of the flask, while the other half lifts off with the cope, or upper part of the flask. It is important to keep the parts directly over each other in the mold to ensure a true casting; this is done by the use of dowel pins, as shown. Split patterns should, whenever possible, be parted and doweled together on the joint or parting of the mold.

Skeleton Patterns

Skeleton patterns are wooden frames that are used in framing a portion of the pattern in sand or clay. After the sand or clay is

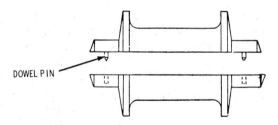

Fig. 10. A typical split, or parted, pattern.

DOWEL PIN

rammed inside the frame, it is worked to the required form by pieces of wood called "strickles."

Part Patterns

Part patterns are sections of a pattern that are so arranged as to form a complete mold by being moved to form each section of the mold. The movements are guided either by following a line or by the use of a central pin or pivot. These patterns are generally applied to circular work.

Patterns With Loose Pieces

Patterns with loose pieces are patterns with projecting parts that form undercuts; these undercuts prevent the loose pieces from being drawn with the main body of the pattern without destroying the mold. The undercutting projections are made as loose pieces and are fastened in place with loose dowels, or "skewers." By removing the skewers, the pattern is freed; the loose pieces which remain in the mold when the pattern is drawn are drawn into the cavity left by the pattern and are removed from the mold; this procedure is called "picking in."

CORES, CORE PRINTS, AND CORE BOXES

Dry sand core is a molded form that is made of sand mixed with a binder and baked until dry and firm. It is placed in the mold to form the interior passage, or opening, in the casting. "Green sand core" is the term used where the sand is allowed to stand in the mold and the metal is permitted to run around it to form the interior passage, or opening, in the casting.

Core Prints

When a casting is to be made hollow, its pattern must be made with a core print, such as the one shown in Fig. 11. This core print leaves a cavity or shelf in the sand mold into which is laid a core, as shown in Fig. 12A. When the metal is poured into the mold, it surrounds the core and leaves an opening of the size and shape of the core in the casting. The sand core is easily knocked out after the metal has cooled.

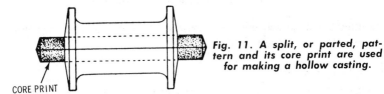

CORE PRINT

Fig. 11. A split, or parted, pattern and its core print are used for making a hollow casting.

There are two kinds of core prints—vertical and horizontal. Core prints should be given a taper in order to more readily draw them from the sand, especially in the case of vertical core prints. The usual core-print taper is approximately ⅛ inch per inch. Horizontal cores must usually be supported at both ends, as shown in Fig. 12A. The length of the core prints at these points is decided by the weight of the core; a heavy core requires longer prints to give sufficient support to the cores and to guard against the crumbling of the sand-mold edges. For vertical cores, the lower core print should be longer than the upper one, since the lower core print practically supports the entire core; the upper core print serves only to keep the core from moving, as shown in Fig. 12B.

Core Boxes

A core is molded to the required shape and size in a core box, Fig. 12C. When both halves of a core are exactly alike, this type of half core box is used to save the patternmaker's time. The core is made in two halves, which are pasted together after they are baked. If it is necessary to make a complete circular core box, the two parts are usually held together by dowels, thus making it easy to remove the core from the box. In a core box for a large pattern, some allowance for expansion should be

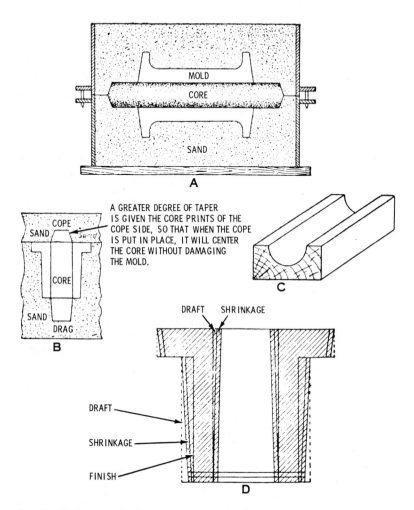

Fig. 12. *Various methods used to obtain a hollow casting; A, horizontal-core method; B, vertical-core method; C, core box; D, draft, shrinkage, and finish allowances required on a pattern for a vertical casting.*

made, since cores expand in baking. If the cores are not dried too quickly, the figure generally used for core-box allowance is $1/16$ inch per foot. To ensure good surfaces on the core, make sure that the face of the core box is smooth.

Draft

A taper, or "draft," must be given to vertical parts of a pattern; if this were not done, it would be impossible to draw the pattern from the molding sand without damaging the mold. This draft is shown in Fig. 12D. The size and shape of a pattern determines the amount of draft to be allowed. However, it is a good practice to allow approximately ⅛ inch per foot of length.

Shrinkage

When the molten metal is poured into a mold, it contracts as it cools and leaves the casting smaller than its mold. Therefore, in order for the casting to be the required size, this shrinkage must be allowed for when making the pattern. The shrinkage allowances usually employed by patternmakers are:

Iron . . . . $1/8$ inch per foot
Steel . . . . $3/16$ inch per foot
Brass . . . . $3/16$ inch per foot

Special rules, called "shrinkage rules," are available for the patternmaker's use. These rules contain the shrinkage of the casting metal, thereby eliminating the necessity of determining the proper allowance for the metal. In appearance, they look the same as an ordinary rule except that they are longer. Thus, a 2-foot shrink rule for ⅛ inch shrinkage per foot will be ¼ inch longer than the standard rule.

Finish

When any of the surfaces of a pattern are to be machined and finished off, an allowance for the finish must be made in addition to that for draft and shrinkage, as shown in Fig. 12D. The amount allowed for this varies from approximately $1/16$ inch to ¼ inch, according to the location and nature of the casting, the methods of machining it, and the degree of finish required.

A good general rule to remember for the allowance on cast iron, small or medium sized work to be finished in a lathe, planer, or milling machine is approximately ⅛ inch per foot. For larger pieces, the allowance will vary from ¼ to ¾ inch. An extra finish allowance is usually made on the cope side of patterns for large castings to permit machining to the sound metal beneath the slag, which always rises to the top of a mold. If a casting is to be finished on only one end, the pattern should be marked so as to make that fact known to the foundry man. Some foundries paint the finished surfaces green.

Blueprints

A blueprint is the plan of a casting from which the patternmaker is required to construct a pattern. On standard blueprints, there is a title block in the lower right-hand corner. It gives the company's name and the draftsman's name. It tells whether the drawing is full size, half size, or quarter size and also describes the kind of material to be used for the casting.

The notes on the blueprint should be carefully studied. When a drawing is to be scaled for dimensions, a standard scale should always be used on the work. If the dimension is marked ¾ *inch*, it means that the drawing is out of scale. If holes are marked *drill*, this means the casting will be drilled. If the blueprint is marked *bored* or with an "*f*" (indicating finish), an allowance should be left on the pattern for this; the amount of the allowance is determined by the kind of material to be used for the casting. On small work where the blueprint is marked *spot face* or *disc grind*, only about $1/_{32}$ inch for the finish allowance is added.

JOINERY

Previous chapters discussed a wide variety of joints. In the patternmaker's art quite a few of them, as well as variations on those joints and other joints, are required. These are presented in the pages that follow. As with any joint, precision is the key. Indeed, without the required precision the pattern simply will not work. While hand tools may be used, the craftsman equipped with stationary power tools will be in a better position to achieve a good result.

The basic joints used in the construction of patterns and core boxes are:

1. Straight or butt.
2. Checked.

3. Half-lapped.
4. Tongue and groove.
5. Splined.
6. Rabbet.
7. Dado.
8. Mortise and tenon.
9. Dovetail.

Butt Joint

The butt joint, Fig. 13A, is the simplest form of joint and is probably used most frequently, especially for light framework. It is not too strong even when properly glued and should be reinforced whenever possible by nails, screws, or dowels. The butt joint is really considered the least effective joint.

Checked Joint

The checked joint, Fig. 13B, is also known as the housed, butt, or rabbeted joint. It has an advantage over the butt joint in that it may be nailed or screwed from the edge. It is also adaptable to the formation of corner fillets.

Half-Lapped Joint

There are three varieties of half-lapped joints in general use—the corner lap, the cross lap, and the center lap. The center lap is sometimes dovetailed, Fig. 13C, D, E, and F. The half-lapped joint is considered the best all-around fastening joint for light frames. When properly glued and fastened with screws, it makes a strong joint that is commonly used in the joining of ribs and webs.

Tongue-and-Groove Joint

The tongue-and-groove joint, Fig. 13G, is often used for fortifying butt joints between ribs, etc., and for plate work in the making of boards for mounting patterns.

Splined Joint

The splined joint in Fig. 13H is the same as the feather joint or the plowed-and-tongued joint. It serves the same purpose as the

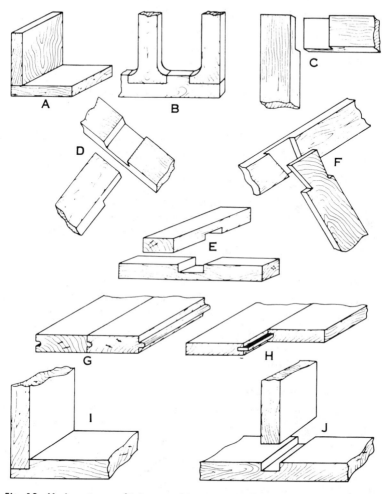

Fig. 13. *Various types of joints used in patternmaking; A, butt; B, checked or housed; C, half-lapped, corner; D, half-lapped, center; E, half-lapped, cross; F, half-lapped, center dovetailed; G, tongue-and-groove; H, splined; I, rabbet; J, dado.*

tongue-and-groove joint, but it is preferred when joining two pieces of softwood because of the added strength given to the joint by the inserted spline, or feather. The spline is usually made of hardwood which is cut across the grain and fitted into grooves in the two joining pieces; this greatly reduces the liability of the joint to snap, as is often the case when the tongue is cut from softwood lengthwise with the grain.

Rabbet Joint

The rabbet, or housed butt, joint, Fig. 13I, is easily made and is satisfactorily used on the sides and ends of core boxes.

Dado Joint

The dado joint, Fig. 13J, is quite satisfactory for fastening the ends of ribs in the sides of frames and boxes.

Mortise-and-Tenon Joint

The mortise-and-tenon joint, though occasionally used with good results in some pattern framework, is scarcely used at all in small pattern work. This joint may be either through, blind, or open, as illustrated in Fig. 14A, B, and C.

Dovetail Joint

The dovetail joint, Fig. 14D, is exceptionally strong and is used in places which are difficult to fasten with screws or nails, such as the corners of light beds and open-sided boxes. It is also used to some extent on loose pieces for first-class permanent patterns and core boxes.

Squaring Framework

To square large frames, use a measuring rod placed diagonally from corner to corner; move the sides of the frame until the two diagonals are equal. On large box work, use the steel square, and check with the rod. Nail a batten to the face to hold it square until the glue sets and also until the corner fillets, if any, have been fastened in place.

Corner Fillets on Thin Framework

On thin frames, corner fillets are frequently made by leaving sufficient stock on the ends of the frame from which they are shaped after the half-lapped joint is made, as shown in Fig. 14E. If the fillets are of large radii, they are usually put in separately because of the cost involved. They are often made of two

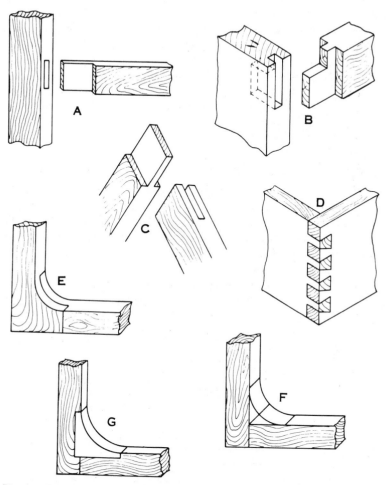

Fig. 14. Various joints used for square and curved patterns; A, mortise and tenon, through; B, mortise and tenon, blind; C, mortise and tenon, open; D, dovetail; E, half-lapped with corner fillets; F, butt with two-piece fillet; G, checked or housed fillet.

pieces, as in Fig. 14F, and sometimes they are made in one piece and are checked or "housed," as in Fig. 14G. They are fastened roughly in place by gluing and nailing and then worked into shape with a paring gouge.

Hardwood Corners and Edges

The corners and edges of pine patterns which are likely to be subjected to hard usage are reinforced with hardwood. Rabbets are cut in the pattern into which the hardwood pieces are fitted; in open boxes, the plowed-and-tongued, or splined, joint is frequently used to fasten the sides and ends to the corners. Ribs are often edged with hardwood. The inside corners of boxed work and the outside corners of core boxes are reinforced with corner blocks of hardwood which are glued and nailed in place.

Joints for Loose Pieces

On high-quality work, loose pieces that are to be left in the mold when the pattern is drawn are dovetailed in place. A dovetail is cut on the pattern, and a piece to fit it (forming the pin) is fastened to the back of the loose piece, as shown in Fig. 15A. When laying out the dovetail, provide an ample taper in the direction of the draw, as shown in Fig. 15B, and plane the pin approximately ½ inch longer and slightly thicker than the depth of the dovetail from the face to the bottom. The additional length allows for cutting off the stock at the bottom if the dovetail should be cut too large, and the added thickness permits the dovetail piece to be planed flush with the pattern face. A

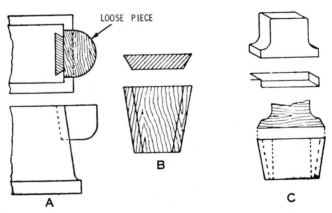

Fig. 15. Joint details for loose pieces; A, the loose piece; B, proper taper, in the direction of the draw, must be provided; C, the loose piece is grooved into the pattern to strengthen the fillet.

backsaw is used to cut the dovetail socket; cut as close to the line as possible and as far as the lines will allow. The stock is removed from the socket with a chisel or finished to depth with a router.

Fillets on Loose Pieces

If fillets are required on loose pieces that are to be "picked in," the loose piece is gained, or grooved, into the pattern, as shown in Fig. 15C, in order to give the fillet a thicker and stronger edge, thus eliminating the feather edge, which is so easily broken. This thick edge is sometimes made in the form of a dovetail, as in the lower section of Fig. 15C

PATTERN DETAILS AND ASSEMBLY

Patternmaker's Box

The patternmaker's box, as shown in Fig. 16, is used as a foundation to make many patterns, especially if they have a regular outline and a rectangular form. The box should be made so as to have as few pieces of end wood as possible in contact with the draw sides. The top and bottom of the box are fitted between the sides and ends to prevent them from extending beyond or falling short of the sides through shrinkage or other causes. To add support to the top and bottom, the ends may be rabbeted, and a central rib may be provided. This rabbet is sometimes formed by extra pieces which are glued and nailed to the inside of the box, as shown.

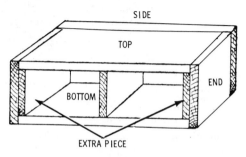

SIDE

TOP

END

BOTTOM

EXTRA PIECE

Fig. 16. The patternmaker's box.

347

Shaping Parts Before Assembling

Occasionally, parts of patterns cannot be worked to shape after the pattern has been put together without great difficulty. In such cases, the parts should be laid out and cut as closely as possible to the lines before gluing. Framework that is to be shaped on the inside, where machine finishing will be difficult after assembly, should be assembled for laying out and then taken apart and sawed. Screws or dogs may be used to fasten the work while it is laid out. Pieces to be fitted in corners should be finished, even to sandpapering, before they are fastened in place.

Fitting to Cylindrical Forms

The usual procedure when fitting to cylindrical pieces is to chalk cylindrical parts and then rub and cut the piece to be fitted until it forms a tight joint. This fitting is usually done on the band saw; some additional fitting afterward may be necessary, but much time should not be taken in rubbing and hand fitting to make a perfect joint, because the fillet covers the joint. When making bosses, use the thickest lumber available, and fit them as far as possible around the cylinder; fill in with one or more pieces after the boss is fastened, as shown in Fig. 17A. Fig. 17B and C illustrates an approved method of fitting a circular boss or branch to a cylindrical body.

Staved or Lagged Work

Staved work is one of three methods used in constructing patterns and core boxes of cylindrical shape. The other two methods are known as stepped work and segment work. A staved, or lagged, pattern is constructed by fastening barrow strips, called "staves" or "lags," to foundation pieces, called "heads." Fig. 18A shows the staves fastened to the heads of a pattern for one-half of a regular cylinder that is to be parted lengthwise through the center. Large cylindrical work that is to be finished either by hand or by turning is usually constructed by this method; it provides the maximum amount of strength and makes it possible to construct close to the finished outline of the pattern.

348

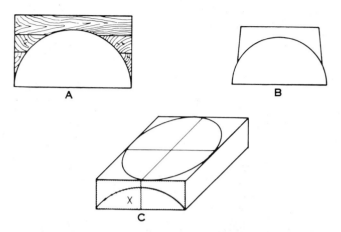

Fig. 17. The method of fitting a circular boss to a cylindrical body. The boss is laid out, as shown, and is sawed on the band saw to fit the cylinder; the waste piece X is used to hold the boss in position on the saw table while the boss is sawed to the required outline.

The staves are ripped in narrow pieces and beveled on each side from stock previously planed to an even thickness; they are then fastened to the heads by means of glue and nails. The number and thickness of staves depend largely on the size of the job (they are usually about 1 inch thick). However, the amount of gluing surface between the staves is important if a good joint is desired; it should not be less than ³/₄ inch.

In medium-size work, the undersides of the staves are often concaved to fit the circle of the head by passing them over the circular saw at an angle. The saw is projected above the table at a distance equal to height X in Fig. 18B, and one of the staves, with the narrow side up, is clamped on the saw table at an angle, as shown in Fig. 18C. This angle is determined by squaring off the stave so that the distance from the edge of the stave to the front of the saw is equal to the width of the narrow side of the stave, as at Y. If the radius required on the stave is greater than the radius of the saw, two or three cuts may be necessary on each stave to cut to the radius line. When a large radius is used, the heads are cut flat under the staves; the circumference is spaced in even parts, and the staves are cut to fit, as in Fig. 18D. A brace, or rapping and lifting bar, is run through each head at

349

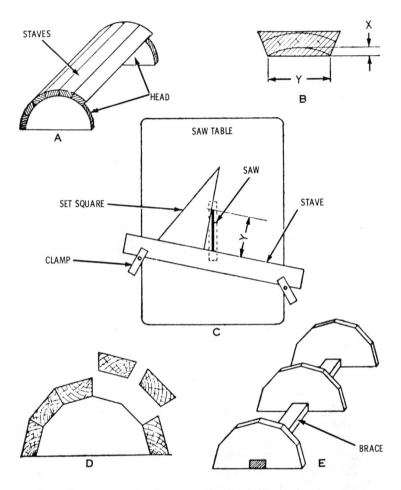

Fig. 18. The staved-work method of making cylindrical patterns; A, staves fastened to the heads of a pattern for half of a regular cylinder; B, cutting the staves; C, staves are clamped and sawed on a saw table; D, staves for a large radius cylinder; E, a brace is used to strengthen the pattern.

the parting line, or center of the pattern, to strengthen the pattern by tying the heads together and to provide the means for rapping and lifting in the foundry, as shown in Fig. 18E.

The assembly of the staves and heads is done on a perfectly

smooth, straight board having a straight edge. A center line is drawn parallel to the straight edge, and lines representing the length of the pattern are squared from this edge across the board. The heads are set to these lines with their center lines matching the center line on the board, as in Fig. 19A; each stave is then glued and nailed in place, beginning at the center or parting line and working to the center or top of each half.

Some patternmakers begin construction by fastening the second stave from the bottom in place first and then continuing until the second stave from the bottom on the opposite side is in place. This is done because the joint staves are left slightly wider than the others, as shown in Fig. 19B, to allow for hand planing of the joint.

If the pattern is to be a complete cylinder, then the second half of each head should be doweled to the first and fastened together with pinch-dogs driven in both sides (inside and outside) of each head, as shown in Fig. 19C. Fasten the joint staves on each side of this half section first, and continue building from these. Before fastening the last stave in place, remove the dogs on the inside of the heads.

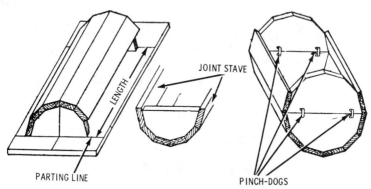

Fig. 19. Assembly of staved patterns.

Staved Core Boxes

The principles involved in the construction of staved core boxes are the same as for staved patterns, only in reverse; the staves are finished on the inside and the outside and are also braced or supported. The ends of staved boxes are called "heads" and are usually connected by braces at the face and bottom, as shown in Fig. 20A. Round heads and concaved staves are used for long boxes, as shown in Fig. 20B. The end heads are depended on to keep the box together, because the outside bracing may or may not be added. The box is built on a board, the same as a cylinder. If the box is exceptionally long, several heads may be used and then removed after the box is put together. This construction produces a true box, which is finished when the staves are put on.

Stepped work gets its name from the fact that the stock, when fastened together, resembles steps. It is the method used for constructing cylindrical forms, such as straight and curved pipes, up to 8 or 9 inches in diameter that are to be finished in the lathe, and for elbows and bends of all sizes that are to be finished by hand.

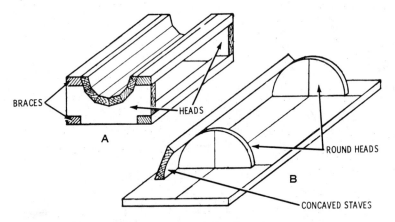

Fig. 20. Assembly of staved core boxes.

Fig. 21 illustrates how a parted cylinder pattern is constructed by the stepped method. A layout is made of one-half of the cylinder and the core print, as shown in Fig. 21A. The lumber used should be planed to thickness, since the number of steps depends on the thickness, and the two top pieces of the stock and the different steps should be cut to width according to the measurements taken on the layout. The outside diameter is divided

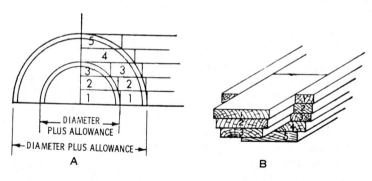

Fig. 21. The construction of a parted-cylinder pattern by the stepped method.

into equal parts on the vertical center line, and horizontal lines are then drawn through these points. Perpendicular lines, which are erected at the intersections of the circles and the horizontal lines, as shown, will give the width of the top pieces and the different steps.

The steps are assembled on the top pieces, as shown in Fig. 21B; the ends are marked so as to locate each piece. They are then glued and held in place with clamps or pinch-dogs. When gluing pieces such as these to a line, toenail each end to prevent the piece from shifting while applying the clamps or pinch-dogs. The pieces for the prints are fitted after the glue has dried. The joint step should be thicker than the others to provide for truing up the face on the jointer.

Segment Work

The primary purpose of segment work is to have the grain of the wood follow the outline of the pattern as far as is possible, because this procedure provides greater strength and makes it easier to finish the pattern, especially if the pattern is to be turned, since most of the end wood is eliminated. Segment work is used for the construction of curved ribs and similar pieces. When a complete circle is to be constructed, successive layers of wood, called "courses," are built up to the thickness of the required piece; these courses are divided into an equal number of divisions, called "segments," as shown in Fig. 22A. The number of courses for a given job depends largely on the thickness of the stock available and also on how this stock corresponds with the total thickness required. At least three rows of courses are necessary to avoid warping. An easy way to determine the number of courses required is to make a layout of a small section of the job, adding the necessary amount of allowance for the finishing operation, and divide this to suit the lumber.

Fig. 22B shows the layout of a segment which is to be used as a pattern to mark the outline on the stock, as shown at **X**, having six segments to each course. When there are six segments to a course, the length of each one is equal to the radius of the circle. An allowance of $1/16$ inch is added to each end of the segment for fitting. The $1/4$-inch allowance is cut off after the segments have been laid up.

The band saw should be used when sawing segments; cut as close to the lines as possible. The ends of the segments are cut to the correct angle on the trimming machine, which is equipped

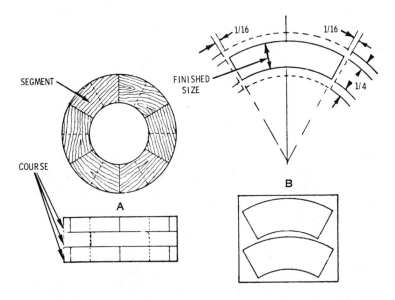

Fig. 22. The method of obtaining a cylindrical pattern by the segment method.

to give the correct angle set for any number of segments in a course from 3 to 12. The different courses in segment work are glued together with the segment joints of one course coming halfway between the joints of the course above and below. Thick glue is used and should be well rubbed into the segment ends to size them. Reinforce the glue by nailing whenever possible.

Core-Box Construction

The common method of making a core box for square or rectangular cores is to have the ends dadoed into the sides, with the box capable of being parted at diagonal corners to release the core, as shown in Fig. 23A. The loose corners are fastened with screws, around the heads of which are drawn circles with black varnish to indicate that they are to be removed to part the box and release the core. The dadoes on the loose corners are usually made more shallow than those on the fastened corners in order to make the box part freely. Corner blocks, securely glued and nailed in place, are used to reinforce the fastened corners. If corner fillets are required, they are glued in place at the fastened corners; at the parting corners, they must be made a part of the side of the box, as shown in Fig. 23B.

Fillers—Blocks known as fillers are placed in rectangular core boxes to make the core shorter or narrower, or both, than the box and to change the outline of the box, as shown in Fig. 24. Fillers are fastened in the box or placed loosely in position as the occasion requires.

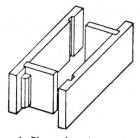

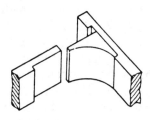

A. Diagonal parting corners.   B. Filleted corners.

Fig. 23. Core-box construction.

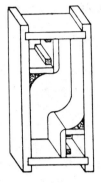

Fig. 24. The use of fillers to make the core shorter, narrower, or both, than the box.

Round Core Boxes—The term "round core box" usually means a half-round box. To lay out a box of this type, plane the stock true on the face and on one edge; square the ends, and cut to length. There should be a margin on the face of the box along each side of the required diameter or not less than 1 inch. A distance equal to this margin is gauged from the working edge; from the pieces bradded to each end of the stock as centers, circles are drawn on each end which coincide with the gauged line on the face, as shown in Fig. 25A. Connect the opposite sides of the circles by a line which has been either gauged or drawn with a straightedge. The circles are laid out on the ends of stepped and lagged boxes by the use of the device shown in Fig. 25B.

Core boxes made from solid timber are roughed out by making a series of cuts with the circular saw up to within ¹/₈ inch of the radius line, as shown in Fig. 25C, and then removing the sawed stock with a gouge. On stepped or lagged boxes, the roughing-out procedure is done with an offset-tang gouge. A core-box plane is used to remove the remaining stock. To start the plane, a rabbet approximately ¹/₁₆ inch deep is planed on both sides of the box by fastening a thin strip of wood to the face of the box along the layout lines as a guide for the plane. Remove the strip and plane from the right side to the center, as in Fig. 25D. Remove the stock, start from the right, and plane to the center again. Sand until smooth with fine sandpaper and a mandrel at least ¹/₁₆ inch smaller than the size of the desired core.

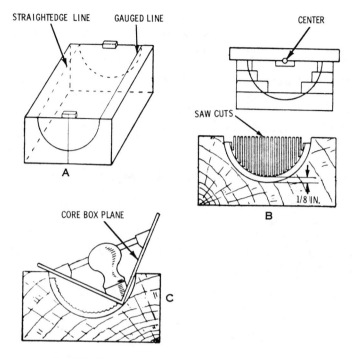

Fig. 25. The construction of round core boxes.

SUMMARY

Wood patternmaking is a most highly skilled branch of the carpentry trade. It takes a great amount of skill and knowledge to develop this specialized work. It requires great skill in making joints with woodworking tools and machinery, in addition to a knowledge of wood carving and turning. A person must master the ability to read blueprints and visualize the shape and form of the pattern from a diagram or blueprint.

A complete set of tools is required in patternmaking, such as various sizes and types of planes, chisels, carving tools, hand and automatic drills, bits, saws, clamps, and squares. Electric machinery such as drills, sanders, saws, and lathes makes it possible to do work rapidly and accurately.

Various cabinetmaking joints, as discussed in Chapter 25, are required in patternmaking, such as butt joints, miter joints, and dadoes. The most common woods used in patternmaking are white pine, mahogany, cherry, maple, and birch.

Staved work is one of three methods used in constructing patterns and core boxes of cylindrical shape. The other two methods are known as stepped work and segment work. A staved pattern is constructed by fastening barrow strips, called staves, to foundation pieces, called heads. The staves are fastened to the heads by means of glue and nails.

REVIEW QUESTIONS

1. What are the most common woods used in patternmaking?
2. What is a patternmaker's box?
3. What is a stave? What is a head?
4. What are the advantages of power-driven machine tools?
5. What is a pinch-dog?

CHAPTER 24

Kitchen Cabinet Construction

Today, many carpenters find it more economical to have kitchen cabinets custom-built by a cabinet shop or to buy stock cabinets directly from these outlets. Still, other carpenters and handymen build their own cabinets from scratch. The information in this chapter will enable you to do this.

The size and number of cabinets required depends on the kitchen area available, the amount of cooking to be done, and the distance of the home from store centers. Thus, a home that is located within a short distance of stores will not need the storage space required in a home that is remotely located from shopping centers, nor would a small family need the same amount of storage space as that required by a large family.

When remodeling a kitchen, certain limitations must be observed. For example, the locations of the sink and range are usually predetermined, thus making it necessary to plan, construct, and install various cabinets and equipment so as to keep these two centers in their original locations.

Once the layout has been decided on, the next step consists of choosing the number and type of cabinets. The cabinets are usually designed so that they can be made as individual units and installed as such, or several cabinets can be made and installed as a combination. When a combination of cabinets is to be arranged to form a large unit, it is usually better to make a large top of sufficient length to cover the two or more units placed side by side than to make a separate top for each unit.

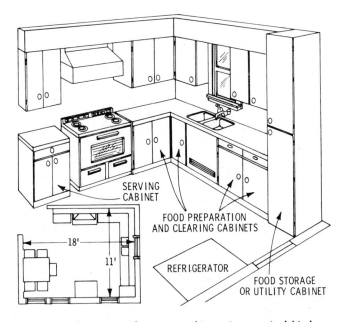

SERVING CABINET

FOOD PREPARATION AND CLEARING CABINETS

18'

11'

REFRIGERATOR

FOOD STORAGE OR UTILITY CABINET

Fig. 1. The arrangement of storage cabinets in a typical kitchen.

The number and the location of electrical outlets and fixtures are other important parts of kitchen planning. Wall outlets should be placed above the floor cabinets for easy connection to electrical appliances used in the preparation of food, such as toasters, food mixers, coffeemakers, casseroles, roasters, and broilers. It is a good idea to have several double outlets available to permit the use of as many appliances as possible. Electric and gas ranges commonly come with permanently installed fluorescent lamps, which prevent shadow problems when working at the kitchen range. To provide kitchen ventilation and to remove cooking odors, an exhaust fan installed in an outside wall is another useful appliance; ranges also, of course, come equipped with vents.

CONSTRUCTION CONSIDERATIONS

Although it may readily be admitted that not every carpenter has sufficient skill to design and make really artistic cabinets suitable for kitchen use, a great many simple cabinets can be easily made. In general, it is best to start out with a cabinet of simple construction, and when greater proficiency has been attained in the handling of tools and the working of wood, the more intricate and difficult projects may be undertaken.

KITCHEN WALL CABINETS

As the name suggests, kitchen wall cabinets are designed to be hung on the wall above the floor cabinets and are suitable companion pieces to the floor units. As detailed in Fig. 2, they are fairly simple in construction and may easily be made with simple tools. The design illustrated permits construction in widths ranging from 16 to 36 inches, depending on the available and desired space.

For the most convenient usage, it has been found that cabinets of this type should be proportioned so that the top of the cabinet will be approximately 7 feet above the floor surface. Cabinets ranging from 16 to 24 inches in width usually require only one door, while cabinets of 28 to 36 inches in width require two doors. It should be noted, however, that regardless of size, all cabinets are similar in construction; the only difference is in the width and height. When the width and height of the cabinet have been decided on, the material for each cabinet should be ordered. As the different pieces are cut, they should be marked with a key letter to facilitate assembly. If the cabinet is to be constructed of board stock rather than plywood or other wide materials, it may be necessary to glue up two or more narrow widths of stock to produce panels having the required widths.

The actual working drawings shown in the accompanying illustrations are based on a cabinet that is 28 inches wide, although the construction procedure will be similar for cabinets of any width.

Construction Details

When making a cabinet such as the one illustrated, it should be noted that actual construction is started with the side members. These pieces, as shown in the illustration, have a rabbet cut on the upper end to take the top piece, a rabbet cut along the

361

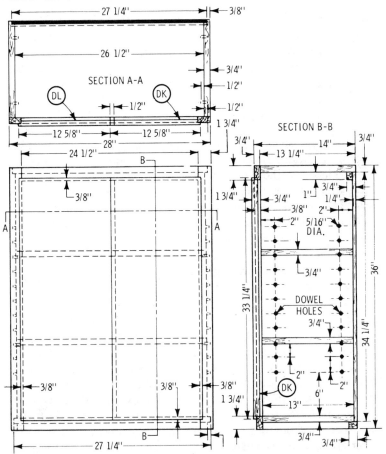

Fig. 2. Construction details of a typical kitchen wall cabinet.

back edge to take the back panel, a dado cut near the lower end for the bottom member, and two grooves cut at the upper and lower back edges to take the back rails.

After the necessary work on the two side members, the carcass for the cabinet can be assembled. When the carcass is completely assembled, the corners should be checked with a try square to make certain that they are square. A temporary diagonal brace may be fastened to the front edge to keep the unit square while the back panel is being applied.

The front frame is constructed of the two stiles and the bottom and top rails. The rails are joined to the stiles by means of mortise and tenon joints. Glue is spread over the joints, and the frame is set in clamps. The corners of the assembled frame must be checked to make certain that they are square. The shelves are cut to size in the regular manner, but they require the cutting of small circular grooves in each end to properly engage the supporting dowels. The door, or doors, should not cause any difficulty when assembling them to the unit if they are properly cut to fit the cabinet.

To complete any cabinet, hardware of some type is required. The types of cabinet hardware available are varied and abundant, as a glance through a hardware catalog shows. Various classes of hinges, door pulls, catches, and latches are generally found at any well-equipped hardware store. If catches are to be placed on double doors, then the strikes should be fastened to the underside of the shelf, to the top rail of the cabinet above the door, to the lower rail below the door, or to the bottom shelf. The strike and catch are usually provided with elongated holes in order to permit their adjustment after they have been installed. If a series of cabinets is being planned, the necessary hardware required may conveniently be purchased at one time, thus assuring the perfect matching of each unit.

Cabinet Installation

Because of the considerable weight of a fully loaded cabinet, such as the combination unit shown in Fig. 3, a great degree of care must be taken to ensure its proper installation. In all homes of frame construction, as well as in many other types, the inside walls are constructed on studs, which are on 16- (usually) or 24-inch centers. These studs will, in most cases, provide the only supporting means for the cabinet. The studs may be located by various means, such as by a fine drill or a thin nail; be careful not to damage the wall during the stud-locating process. A plumb line may be dropped from the ceiling at the center of the studs to obtain the center line of the studs throughout their entire length. Once the location has been determined, it is a simple matter to locate the holes in the top and bottom rails of the cabinet for final installation by means of suitable wood screws.

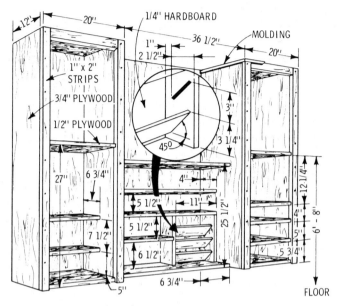

Fig. 3. *The layout of a wall-type kitchen-cabinet unit with center canned-food racks. Metal angles on the wall are used to support the cabinet.*

Use of a False Cabinet Wall

Most kitchen wall cabinets are designed so that the hanging wall units and full-length cabinets have a maximum height of 7 feet above the floor. The reason for observing this standard is for easier use, since any shelf placed higher than this distance cannot normally be reached without the aid of a stepladder. With cabinets installed at standard heights, however, there will usually be an open space near the ceiling where dust will tend to accumulate. A convenient method of eliminating this dust-collecting space consists of constructing a false wall extending from the top of the cabinet to the ceiling. This method, when used, will add to the decor of the room and is adaptable to one or several cabinets as required.

To enclose the space above the cabinets, a framework, such as shown in Fig. 4 must be installed. Sheetrock or plywood can then be attached to this framework. The stock used for the framework need not be of heavier cross section than $1^{1}/_{4}''$ ×

$2^1/_2''$ or even lighter, depending on the bracing and height of the upright members. As noted in Fig. 4, cleats are fastened to the cabinet top and the ceiling to provide nailing strips for the uprights. Since the cleats are notched to receive the uprights, the panels may be fastened flush with the framework on all sides. Before nailing the cleats in place, however, the thickness of the panel must be determined, since the setback of the cleats which are attached to the cabinet top is controlled by this measurement. After having properly located the upper cleat, the next step is to locate the ceiling beams. If the ceiling beams run at right angles to the cleat, no difficulty will be encountered when fastening the cleat to the ceiling. If, on the other hand, the beams are found to run parallel to the cleat, the nailing strips will have to be applied at right angles to the beams and extended out far enough to provide a nailing surface for the cleats.

After the framework has been nailed in place, all that is necessary is to cut the panels to size. These are usually fastened by means of four-penny nails. Scrollwork, when desired, is an op-

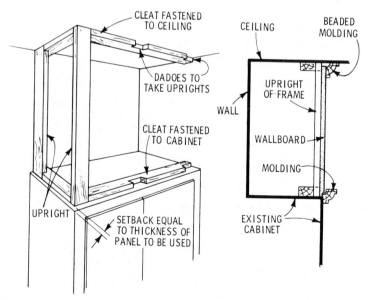

Fig. 4. A simple construction method for enclosing the wall space above an existing wall cabinet.

tional feature. When applying the molding to the false cabinet wall, care should be taken to match the existing molding as closely as possible. These panels can usually be purchased at any well-equipped lumberyard.

The next step is the sanding and finishing operations. The sanding removes whatever scratches are left in the wood; the false cabinets are then finished to match the original cabinets. Also, if the paneled surface is to be papered, this work must be done before the molding is fastened.

BASE CABINET CONSTRUCTION

A typical kitchen cabinet of the base type is shown in Fig. 5. The cabinet is 24 inches deep from front to back and is 36 inches high. The cabinet, in most cases, is constructed of ³/₄ inch plywood, although particleboard and flake board are sometimes used. The rails and stiles are usually cut from ³/₄ inch plywood, although solid stock can be used. The sides of the base suit are dadoed to receive the shelf and plywood bottom. A 3¹/₂ ×

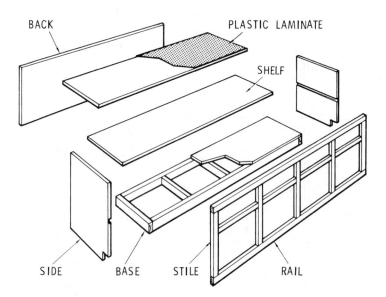

Fig. 5. Construction details of a typical kitchen unit base cabinet.

$3^1/_2$-inch toe-space is provided at the bottom of the unit and two-by-fours are placed under the plywood bottom. The sides of the base units are sometimes rabbeted to receive a $^1/_4$-inch hardboard back or a 1 × 6 hanging strip is used in place of backing. All joints should be glued, using brads as required to anchor the work and prevent slipping.

Doors and Drawers

There are two basic types of drawers and doors that can be used, flush and lipped. The flush door fits between the rail and stile and is very difficult to fit. The lipped door has a $^3/_8$-inch rabbet around the door, allowing for a margin of error. If a lipped door or drawer is used, the size of the front of the drawer or door is obtained by adding $^1/_2$ inch to the size of the opening.

Plastic Laminate Tops

The standard countertop these days is composed of a base of particleboard or plywood covered by plastic laminate, a high-pressure, extremely durable, and hard material that resists moisture, scratches, and various other assaults.

Plastic laminate is available in a wide variety of colors, textures, and patterns, everything from plain colors to simulated wood; textures may resemble many materials, including stone.

Installation of plastic laminate is relatively simple using contact cement, but unless you use the more expensive water-based cement to adhere it, the job must be done outdoors. Regular contact cement gives off fumes that are highly combustible, even explosive.

Plastic laminate, which is $^1/_{16}$ inch thick and comes in large sheets, may be cut with a special cutter, hacksaw, fine-toothed saw, or coping saw. But the easiest way to cut it is with a router equipped with a laminate cutting bit.

There are a number of ways you can secure the laminate. A good procedure is as follows: Cut the laminate about an inch oversize, all around, then the base material—plywood or particleboard. Use a brush or roller to apply contact cement to both surfaces. Make sure every square inch is covered. When the laminate is just dry to the touch—a piece of paper will not stick to it—position it over the base material and lay a series of sticks across the material. Lay the laminate in place and then, one by one, withdraw the sticks. As you do so, press the laminate in place, pounding it down with the heel of your hand so it sticks securely. When all the sticks have been withdrawn, use a wood block to pound it down further. Secure the edge strips, also trimming these to fit once they are secured. Finally, use a wood block and a hammer to pound the laminate down. It's also a good idea to use a protective pad between the block and the laminate.

SUMMARY

The general number of cabinets required for the kitchen area depends on the size of the kitchen and the amount of cooking to be done. When remodeling a kitchen, there are certain limitations to observe; for example, the location of the sink and cooking range. The number and location of electrical outlets and fixtures are other important elements in kitchen planning.

Because of the considerable weight of a fully loaded cabinet, a great degree of care must be taken to ensure proper fit at all joints. Most kitchen wall cabinets are designed so that the hanging wall units and full-length cabinets have a maximum height of 7 feet above the floor.

In most cases plastic laminate material is used for countertops. This material is generally $1/16$ inch thick, resulting in a smooth surface. New countertops should be constructed of $3/4$-inch exterior-grade plywood or particleboard, which should be firmly fastened to the cabinet.

To complete any cabinet, hardware of some type is required. Various classes of hinges, door pulls, catches, and latches are generally found at any well-equipped hardware store.

REVIEW QUESTIONS

1. What determines the number of cabinets and the general placement?
2. What type material is installed on countertops?
3. What type of grade of lumber is generally used in cabinet making?
4. What is generally done to the wall space above the wall-mounted cabinets?

Index

A

Adze, 162
Annual rings, wood, 10, 11
Application of the square, 229
Ash wood, properties of, 17
Augers, 189
Awls, 102, 189
Axe, 161

B

Back
  saw, 128
  steel square, 86
Bench,
  hook, 72
  stop, 70
  support pegs, 68
Beveled joints, 301, 315
Bevel of cutting edge, plane, 179
Black walnut wood, 20
Blind dovetail joints, 294
Block plane, 171
Blueprints, 341
Body, steel square, 227
Bolts, 57-65
  kinds of, 57
  proportions and strength, 60
Boring tools, 189-199
  auger, 189
  brace and bit, 196
  countersink, 195
  hand drills, 196
  scratch awl, 189
  twist drills, 195

Box
  -end wrench, 209
  miter, 92
Brace and bit, 196
Brads, nails, 29
Brown ash wood, 17
Butt
  chisel, 152
  joint, 270, 342

C

Cabinetmaking joints, 299-323
Caring and use, chisel, 155
Carpenters
  pencil, 100
  rule, 107
    folding, 108, 109
  tools, 75-78
Chalk box and line, 100
Chisels, 149
  butt, 152
  firmer, 150
  gouge, 151
  mill, 152
  paring, 150
  pocket, 152
  tang and socket, 151
Clamps, 118
Classifications of wood, 9-11
Combination square, 82, 89
Compass
  and divider, 103
  or keyhole saw, 129
Construction with steel square,
  rafters

INDEX

cripple, 246
hip, 243
jack, 245
valley, 244
Coping saw, 128
Core-box construction, 355
Countersinks, 195
Crosscut saw, 130
Cypress wood, 18

D

Dado joint, 278, 344
Decay of wood, 21
Defects in wood, 13
Dimension of screws, 46
Doors and drawers, cabinet, 367
Double iron, plane, 180
Douglas fir wood, 20
Dovetail joints, 290, 314, 344
Dowel
  joints, 271, 305
  pins, 331
Drawknife, 157
Dressing
  handsaw, 147
Driving nails, 39

E

Eastern red cedar wood, 18

F

Fastening tools, 201-213
  hammer, 201
  screwdriver, 205
  wrenches, 208
Files and rasps, 134
Filing, handsaw, 143
Finding rafter length without table,
  246
Firmer chisel, 150
Fitting cylindrical forms, 348
Fore plane, 170
Framing
  joint, 321
  square, 85

G

Glued joints, 300
Gouge chisel, 151

Grinding
  tools, 219
  wheel, 215
Grooving plane, 173
Guiding and testing tools, 79-98
  level, 95
  miter box, 92
  plumb bob, 95
  square, 80
    center, 91
    combination, 89
    framing, 85
    miter, 82
    sliding-T, 91
    steel, 85
    try, 81
  straightedge, 79
Gum red wood, 18

H

Hacksaw, 128, 129
Half-lapped joint, 342
Halved and bridle joint, 307
Hammer, 201
Hand
  axe, 161
  drills, 196
Handsaw sharpening, 139
  dressing, 147
  filing, 143
  jointing, 139
  setting, 142
  shaping, 140
Hardboard, 23
Hard maple wood, 19
Hatchet, 161
Hemlock wood,
  eastern, 19
  western, 19
Hickory wood, 19
Holding
  tools, 117
    clamps, 118
    horses, 117
    trestles, 117
    vises, 68, 121
Honing, tools, 221
Hook, workbench, 72
Horses, holding tool, 117

How to
  sharpen tools, 219-225
    grinding, 219
    honing, 221
  use a plane, 181
  use a steel square, 227-267
    application of, 229
    constructing with, 243-246
    finding rafter length, 246
    rafter table, 251
    reading rafter length, 257

I

Irons or cutters, plane, 175

J

Jack plane, 169
Jointer plane, 171
Jointing, handsaw, 139
Joints and joinery, 269-297, 341
  cabinetmaking, 299-323
  types of
    beveled, 301, 315
    blind dovetail, 294
    butt, 270, 342
    dado, 278, 344
    dovetail, 290, 314, 344
    dowel, 271, 305
    framing, 321
    glued, 300
    half-lapped, 342
    halved and bridle, 307
    lap, 277
    lap and half-blind dovetail, 294
    mitered corner, 273, 319
    mortise and tenon, 282, 308, 344
    plain, 270
    plowed and splined, 301
    rabbetted, 278, 344
    scarf, 279
    splice, 277
    splined, 276, 342
    square corner, 272, 344
    tongue and grooved, 296, 342

K

Kinds of nails, 28
Kitchen cabinet construction, 359-368
  construction details, 361
  wall cabinets, 361

L

Lag screws, 51
Lap and half-blind dovetail joint, 294
Lap joints, 277
Level, 95
Locust, black wood, 19
Lumber scales, 110

M

Maple wood
  hard, 19
  soft, 19
Marking
  gauges, 111
    double-bar, 112
    single-bar, 112
    slide, 113
  tools, 99-105
    carpenter's pencil, 100
    chalk box and line, 100
    compass and divider, 103
    scratch awl, 102
    scriber, 103
Materials, patternmaking, 329
Measuring tools, 107-116
  carpenter's, rule, 107
    various folding, 108, 109
  lumber scales, 110
  marking gauges, 111
    double-bar, 112
    single-bar, 112
    slide, 113
Mill chisel, 152
Miter
  box, 92
  square, 82
Mitered corner joints, 273, 319
Modified forms of wood, 21
Monkey wrench, 210, 211
Mortise and tenon joints, 282, 308, 344
Mouth, plane, 180

N

Nails, 25-43
  driving, 39
  kinds of,
    brads, 29
    spikes, 30
    tacks, 29

INDEX

power of nails, 36
selection of, 39

O

Oak, red wood, 19
Oilstones, 217
Open-end wrench, 209

P

Paring chisel, 150
Particle board, 23
Pattern
  colors, 331
  details and assembly, 347
Patternmaker's box, 347
Patternmaking, 325-358
  materials, 329
  tools, 326
  trade terms, 328
Pine wood
  long-leaf Southern, 19
  red or Norway, 19
  short-leaf Southern, 20
  white, 19
Plain joints, 270
Plane, 169
  bevel of cutting edge, 179
  double irons, 180
  how to use, 181
  irons or cutters, 175
  mouth, 180
  sharpening, 181
  types of,
    block, 171
    fore, 170
    grooving, 173
    jack, 169
    jointer, 171
    rabbet, 172
    router, 173
    smoothing, 171
Plastic laminate top, 367
Plowed-and-splined joints, 301
Plumb bob, 95
Plywood, 21
Pocket chisel, 152
Poplar, yellow wood, 20

R

Rabbet plane, 172
Rabbetted joints, 278, 344

Rafter tables, 251
Rasps and files, 134
Reading total length of rafter, 251
Redwood, 20
Rip saw, 131
Rough facing tools, 161-165
  adze, 163
  axe, 161
    hand, 163
    hatchet, 161
Router plane, 173

S

Sandpaper, 135
Saw, 127
  back, 128
  compass or keyhole, 129
  coping, 128
  crosscut, 130
  hacksaw, 128, 129
  rip, 131
  sharpening, 139-147
    dressing, 147
    filing, 143
    jointing, 139
    setting, 142
    shaping, 140
  teeth, 129
    angle, 131
    set, 129
Scale problems, steel square, 229
Scarf joints, 279
Scrapers, 185
Scratch awl, 102, 189
Screwdrivers, 205
Screws, 45-55
  dimensions of, 46
  lag, 51
  shapes of heads, 46
  strength of, 49
Scriber, 103
Segment work, 353
Selecting a wood chisel, 153
Selection of lumber, 16
Setting, handsaw, 142
Shape of heads, screws, 46
Shaping, handsaw, 140
Sharp-edged cutting tools, 149-160
  chisels, 149
    butt, 152

firmer, 150
  gouge, 151
  mill, 152
  paring, 150
  pocket, 152
  tang and socket, 151
drawknife, 157
Sharpening
  oilstones, 217
  a plane, 181
  saws, 139
    dressing, 147
    filing, 143
    jointing, 139
    setting, 142
    shaping, 140
  tools, 215-225
    grinding wheels, 215
Sharpen wood chisel, 155
Shellac, 331
Sliding T-square, 91
Smooth facing tools, 167-188
  planes, 169
  scrapers, 185
  spokeshave, 167
Smoothing plane, 171
Socket wrench, 210
Soft maple wood, 19
Spikes, nails, 30
Splined joints, 276, 342
Spokeshave, 167
Spruce wood,
  eastern, 20
  Engelmann, 20
  Sitka, 20
Square, 80
  -and-bevel problems, 235
  combination, 82, 89
  corner joints, 272
  framing, 85
  miter, 82, 92
  sliding-T, 91
  steel, 85
  try, 81
Staved
  and lagged work, 348
  core boxes, 351
Steel square, 85
Stop, workbench, 70
Straightedge, 79
Support pegs, workbench, 68

Surform tools, 135

T

Table problems, steel square, 239
Tacks, nails, 29
Tamarack wood, 20
Tang and socket chisel, 151
Teeth, saw, 129, 131
Testing tools, 79-98
The "penny" system, nails, 27
Tongue and groove joint, 296, 342
Tool panel, workbench, 73
Toothed cutting tools, 127-138
  files, 134
  sandpaper, 135
  saws, 127
    back, 128
    compass or keyhole, 129
    coping, 128
    crosscut, 130
    hacksaw, 128, 129
    rip, 131
Trestles, holding tool, 117
Try square, 81
Twist drill, 195
Types of pattern, cabinetmaking, 335

U

Use of false cabinet wall, 364

V

Varnish, 331
Vises, 68, 121

W

Wall cabinets, 361
Walnut wood,
  black, 20
  white, 21
Western red cedar wood, 18
White
  cedar wood, 18
  oak wood, 19
  or gray ash wood, 17
Withdrawal force of nails, 35, 36
Wood patternmaking, 325-358
Woods, 9-24
  classification, 9-11
  decay of, 21

INDEX

defects, 13
modified forms, 21
selection of, 16
Wood screw head diameter, 46
Workbench, 67-73
    attachments, 67
    bench
        hook, 72
        stop, 70
    support pegs, 68
    tool panel, 73

Wrenches, 208
    box-end, 209
    monkey, 210, 211
    open-end, 209
    socket, 210

Y

Yellow poplar wood, 20

**Over a Century of Excellence
for the Professional
and
Vocational Trades and the Crafts**

**Order now from your local bookstore
or use the convenient order form at
the back of this book.**

AUDEL

These fully illustrated, up-to-date guides and manuals mean a better job done for mechanics, engineers, electricians, plumbers, carpenters, and all skilled workers.

Contents

Electrical II
Machine Shop and Mechanical Trades III
Plumbing IV
Heating, Ventilating and Air Conditioning IV
Pneumatics and Hydraulics V
Carpentry and Construction V
Woodworking VI
Maintenance and Repair VI
Automotive and Engines VII
Drafting VII
Hobbies VII
Macmillan Practical Arts Library VIII

Electrical

House Wiring sixth edition
Roland E. Palmquist
5½ x 8¼ Hardcover 256 pp. 150 illus.
ISBN: 0-672-23404-1 $13.95

Rules and regulations of the current National Electrical Code® for residential wiring, fully explained and illustrated: • basis for load calculations • calculations for dwellings • services • nonmetallic-sheathed cable • underground feeder and branch-circuit cable • metal-clad cable • circuits required for dwellings • boxes and fittings • receptacle spacing • mobile homes • wiring for electric house heating.

Practical Electricity fourth edition
Robert G. Middleton; revised by L. Donald Meyers
5½ x 8¼ Hardcover 504 pp. 335 illus.
ISBN: 0-672-23375-4 $14.95

Complete, concise handbook on the principles of electricity and their practical application: • magnetism and electricity • conductors and insulators • circuits • electromagnetic induction • alternating current • electric lighting and lighting calculations • basic house wiring • electric heating • generating stations and substations.

Guide to the 1984 Electrical Code®
Roland E. Palmquist
5½ × 8¼ Hardcover 664 pp. 225 illus.
ISBN: 0-672-23398-3 $19.95

Authoritative guide to the National Electrical Code® for all electricians, contractors, inspectors, and homeowners: • terms and regulations for wiring design and protection • wiring methods and materials • equipment for general use • special occupancies • special equipment and conditions • and communication systems. Guide to the 1987 NEC® will be available in mid-1987.

Mathematics for Electricians and Electronics Technicians
Rex Miller
5½ x 8¼ Hardcover 312 pp. 115 illus.
ISBN: 0-8161-1700-4 $14.95

Mathematical concepts, formulas, and problem solving in electricity and electronics: • resistors and resistance • circuits • meters • alternating current and inductance • alternating current and capacitance • impedance and phase angles • resonance in circuits • special-purpose circuits. Includes mathematical problems and solutions.

Fractional Horsepower Electric Motors
Rex Miller and Mark Richard Miller
5½ x 8¼ Hardcover 436 pp. 285 illus.
ISBN: 0-672-23410-6 $15.95

Fully illustrated guide to small-to-moderate-size electric motors in home appliances and industrial equipment: • terminology • repair tools and supplies • small DC and universal motors • split-phase, capacitor-start, shaded pole, and special motors • commutators and brushes • shafts and bearings • switches and relays • armatures • stators • modification and replacement of motors.

Electric Motors
Edwin P. Anderson; revised by Rex Miller
5½ x 8¼ Hardcover 656 pp. 405 illus.
ISBN: 0-672-23376-2 $14.95

Complete guide to installation, maintenance, and repair of all types of electric motors: • AC generators • synchronous motors • squirrel-cage motors • wound rotor motors • DC motors • fractional-horsepower motors • magnetic contractors • motor testing and maintenance • motor calculations • meters • wiring diagrams • armature windings • DC armature rewinding procedure • and stator and coil winding.

Home Appliance Servicing fourth edition
Edwin P. Anderson; revised by Rex Miller
5½ x 8¼ Hardcover 640 pp. 345 illus.
ISBN: 0-672-23379-7 $15.95

Step-by-step illustrated instruction on all types of household appliances: • irons • toasters • roasters and broilers • electric coffee makers • space heaters • water heaters • electric ranges and microwave ovens • mixers and blenders • fans and blowers • vacuum cleaners and floor polishers • washers and dryers • dishwashers and garbage disposals • refrigerators • air conditioners and dehumidifiers.

Television Service Manual
fifth edition
Robert G. Middleton; revised by
Joseph G. Barrile
5½ x 8¼ Hardcover 512 pp. 395 illus.
ISBN: 0-672-23395-9 $15.95

Practical up-to-date guide to all aspects of television transmission and reception, for both black and white and color receivers: • step-by-step maintenance and repair • broadcasting • transmission • receivers • antennas and transmission lines • interference • RF tuners • the video channel • circuits • power supplies • alignment • test equipment.

Electrical Course for Apprentices and Journeymen
second edition
Roland E. Palmquist
5½ x 8¼ Hardcover 478 pp. 290 illus.
ISBN:0-672-23393-2 $14.95

Practical course on operational theory and applications for training and re-training in school or on the job: • electricity and matter • units and definitions • electrical symbols • magnets and magnetic fields • capacitors • resistance • electromagnetism • instruments and measurements • alternating currents • DC generators • circuits • transformers • motors • grounding and ground testing.

Questions and Answers for Electricians Examinations eighth edition
Roland E. Palmquist
5½ x 8¼ Hardcover 320 pp. 110 illus.
ISBN: 0-672-23399-1 $12.95

Based on the current National Electrical Code®, a review of exams for apprentice, journeyman, and master, with explanations of principles underlying each test subject: • Ohm's Law and other formulas • power and power factors • lighting • branch circuits and feeders • transformer principles and connections • wiring • batteries and rectification • voltage generation • motors • ground and ground testing.

Machine Shop and Mechanical Trades

Machinists Library
fourth edition 3 vols
Rex Miller
5½ x 8¼ Hardcover 1,352 pp. 1,120 illus.
ISBN: 0-672-23380-0 $38.95

Indispensable three-volume reference for machinists, tool and die makers, machine operators, metal workers, and those with home workshops.

Volume I, Basic Machine Shop
5½ x 8¼ Hardcover 392 pp. 375 illus.
ISBN: 0-672-23381-9 $14.95

• Blueprint reading • benchwork • layout and measurement • sheet-metal hand tools and machines • cutting tools • drills • reamers • taps • threading dies • milling machine cutters, arbors, collets, and adapters.

Volume II, Machine Shop
5½ x 8¼ Hardcover 528 pp. 445 illus
ISBN: 0-672-23382-7 $14.95

• Power saws • machine tool operations • drilling machines • boring • lathes • automatic screw machine • milling • metal spinning.

Volume III, Toolmakers Handy Book
5½ x 8¼ Hardcover 432 pp. 300 illus.
ISBN: 0-672-23383-5 $14.95

• Layout work • jigs and fixtures • gears and gear cutting • dies and diemaking • toolmaking operations • heat-treating furnaces • induction heating • furnace brazing • cold-treating process.

Mathematics for Mechanical Technicians and Technologists
John D. Bies
5½ x 8¼ Hardcover 392 pp. 190 illus.
ISBN: 0-02-510620-1 $17.95

Practical sourcebook of concepts, formulas, and problem solving in industrial and mechanical technology: • basic and complex mechanics • strength of materials • fluidics • cams and gears • machine elements • machining operations • management controls • economics in machining • facility and human resources management.

Millwrights and Mechanics Guide
third edition
Carl A. Nelson
5½ x 8¼ Hardcover 1,040 pp. 880 illus.
ISBN: 0-672-23373-8 $22.95

Most comprehensive and authoritative guide available for millwrights and mechanics at all levels of work or supervision: • drawing and sketching

• machinery and equipment installation • principles of mechanical power transmission • V-belt drives • flat belts • gears • chain drives • couplings • bearings • structural steel • screw threads • mechanical fasteners • pipe fittings and valves • carpentry • sheet-metal work • blacksmithing • rigging • electricity • welding • pumps • portable power tools • mensuration and mechanical calculations.

Welders Guide third edition
James E. Brumbaugh
5½ x 8¼ Hardcover 960 pp. 615 illus.
ISBN: 0-672-23374-6 $23.95

Practical, concise manual on theory, operation, and maintenance of all welding machines: • gas welding equipment, supplies, and process • arc welding equipment, supplies, and process • TIG and MIG welding • submerged-arc and other shielded-arc welding processes • resistance, thermit, and stud welding • solders and soldering • brazing and braze welding • welding plastics • safety and health measures • symbols and definitions • testing and inspecting welds. Terminology and definitions as standardized by American Welding Society.

Welder/Fitters Guide
John P. Stewart
8½ x 11 Paperback 160 pp. 195 illus.
ISBN: 0-672-23325-8 $7.95

Step-by-step instruction for welder/fitters during training or on the job: • basic assembly tools and aids • improving blueprint reading skills • marking and alignment techniques • using basic tools • simple work practices • guide to fabricating weldments • avoiding mistakes • exercises in blueprint reading • clamping devices • introduction to using hydraulic jacks • safety in weld fabrication plants • common welding shop terms.

Sheet Metal Work
John D. Bies
5½ x 8¼ Hardcover 456 pp. 215 illus.
ISBN: 0-8161-1706-3 $17.95

On-the-job sheet metal guide for manufacturing, construction, and home workshops: • mathematics for sheet metal work • principles of drafting • concepts of sheet metal drawing • sheet metal standards, specifications, and materials • safety practices • layout • shear cutting • holes • bending and folding • forming operations • notching and clipping • metal spinning • mechanical fastening • soldering and brazing • welding • surface preparation and finishes • production processes.

Power Plant Engineers Guide

third edition

Frank D. Graham; revised by Charlie Buffington

5$\frac{1}{2}$ x 8$\frac{1}{4}$ Hardcover 960 pp. 530 illus.
ISBN: 0-672-23329-0 $16.95

All-inclusive question-and-answer guide to steam and diesel-power engines: • fuels • heat • combustion • types of boilers • shell or fire-tube boiler construction • strength of boiler materials • boiler calculations • boiler fixtures, fittings, and attachments • boiler feed pumps • condensers • cooling ponds and cooling towers • boiler installation, startup, operation, maintenance and repair • oil, gas, and waste-fuel burners • steam turbines • air compressors • plant safety.

Mechanical Trades Pocket Manual

second edition

Carl A. Nelson

4 × 6 Paperback 364 pp. 255 illus.
ISBN: 0-672-23378-9 $10.95

Comprehensive handbook of essentials, pocket-sized to fit in the tool box: • mechanical and isometric drawing • machinery installation and assembly • belts • drives • gears • couplings • screw threads • mechanical fasteners • packing and seals • bearings • portable power tools • welding • rigging • piping • automatic sprinkler systems • carpentry • stair layout • electricity • shop geometry and trigonometry.

Plumbing

Plumbers and Pipe Fitters Library

third edition 3 vols

Charles N. McConnell; revised by Tom Philbin

5$\frac{1}{2}$x8$\frac{1}{4}$ Hardcover 952 pp. 560 illus.
ISBN: 0-672-23384-3 $34.95

Comprehensive three-volume set with up-to-date information for master plumbers, journeymen, apprentices, engineers, and those in building trades.

Volume 1, Materials, Tools, Roughing-In
5$\frac{1}{2}$ x 8$\frac{1}{4}$ Hardcover 304 pp. 240 illus.
ISBN: 0-672-23385-1 $12.95

• Materials • tools • pipe fitting • pipe joints • blueprints • fixtures • valves and faucets.

Volume 2, Welding, Heating, Air Conditioning
5$\frac{1}{2}$ x 8$\frac{1}{4}$ Hardcover 384 pp. 220 illus.
ISBN: 0-672-23386-X $13.95

• Brazing and welding • planning a heating system • steam heating systems • hot water heating systems • boiler fittings • fuel-oil tank installation • gas piping • air conditioning.

Volume 3, Water Supply, Drainage, Calculations
5$\frac{1}{2}$ x 8$\frac{1}{4}$ Hardcover 264 pp. 100 illus.
ISBN: 0-672-23387-8 $12.95

• Drainage and venting • sewage disposal • soldering • lead work • mathematics and physics for plumbers and pipe fitters.

Home Plumbing Handbook

third edition

Charles N. McConnell

8$\frac{1}{2}$ x 11 Paperback 200 pp. 100 illus.
ISBN: 0-672-23413-0 $10.95

Clear, concise, up-to-date fully illustrated guide to home plumbing installation and repair: • repairing and replacing faucets • repairing toilet tanks • repairing a trip-lever bath drain • dealing with stopped-up drains • working with copper tubing • measuring and cutting pipe • PVC and CPVC pipe and fittings • installing a garbage disposals • replacing dishwashers • repairing and replacing water heaters • installing or resetting toilets • caulking around plumbing fixtures and tile • water conditioning • working with cast-iron soil pipe • septic tanks and disposal fields • private water systems.

The Plumbers Handbook

seventh edition

Joseph P. Almond, Sr.

4 × 6 Paperback 352 pp. 170 illus.
ISBN: 0-672-23419-X $10.95

Comprehensive, handy guide for plumbers, pipe fitters, and apprentices that fits in the tool box or pocket: • plumbing tools • how to read blueprints • heating systems • water supply • fixtures, valves, and fittings • working drawings • roughing and repair • outside sewage lift station • pipes and pipelines • vents, drain lines, and septic systems • lead work • silver brazing and soft soldering • plumbing systems • abbreviations, definitions, symbols, and formulas.

Questions and Answers for Plumbers Examinations

second edition

Jules Oravetz

5$\frac{1}{2}$ x 8$\frac{1}{4}$ Paperback 256 pp. 145 illus.
ISBN: 0-8161-1703-9 $9.95

Practical, fully illustrated study guide to licensing exams for apprentice, journeyman, or master plumber: • definitions, specifications, and regulations set by National Bureau of Standards and by various state codes

• basic plumbing installation • drawings and typical plumbing system layout • mathematics • materials and fittings • joints and connections • traps, cleanouts, and backwater valves • fixtures • drainage, vents, and vent piping • water supply and distribution • plastic pipe and fittings • steam and hot water heating.

HVAC

Air Conditioning: Home and Commercial

second edition

Edwin P. Anderson; revised by Rex Miller

5$\frac{1}{2}$ x 8$\frac{1}{4}$ Hardcover 528 pp. 180 illus.
ISBN: 0-672-23397-5 $15.95

Complete guide to construction, installation, operation, maintenance, and repair of home, commercial, and industrial air conditioning systems, with troubleshooting charts: • heat leakage • ventilation requirements • room air conditioners • refrigerants • compressors • condensing equipment • evaporators • water-cooling systems • central air conditioning • automobile air conditioning • motors and motor control.

Heating, Ventilating and Air Conditioning Library

second edition 3 vols

James E. Brumbaugh

5$\frac{1}{2}$ x 8$\frac{1}{4}$ Hardcover 1,840 pp. 1,275 illus.
ISBN: 0-672-23388-6 $42.95

Authoritative three-volume reference for those who install, operate, maintain, and repair HVAC equipment commercially, industrially, or at home. Each volume fully illustrated with photographs, drawings, tables and charts.

Volume I, Heating Fundamentals, Furnaces, Boilers, Boiler Conversions
5$\frac{1}{2}$ x 8$\frac{1}{4}$ Hardcover 656 pp. 405 illus.
ISBN: 0-672-23389-4 $16.95

• Insulation principles • heating calculations • fuels • warm-air, hot water, steam, and electrical heating systems • gas-fired, oil-fired, coal-fired, and electric-fired furnaces • boilers and boiler fittings • boiler and furnace conversion.

Volume II, Oil, Gas and Coal Burners, Controls, Ducts, Piping, Valves
5$\frac{1}{2}$ x 8$\frac{1}{4}$ Hardcover 592 pp. 455 illus.
ISBN: 0-672-23390-8 $15.95

• Coal firing methods • thermostats and humidistats • gas and oil controls and other automatic controls •

ducts and duct systems • pipes, pipe fittings, and piping details • valves and valve installation • steam and hot-water line controls.

Volume III, Radiant Heating, Water Heaters, Ventilation, Air Conditioning, Heat Pumps, Air Cleaners
5 1/2 x 8 1/4   Hardcover   592 pp.   415 illus.
ISBN: 0-672-23391-6 $14.95

• Radiators, convectors, and unit heaters • fireplaces, stoves, and chimneys • ventilation principles • fan selection and operation • air conditioning equipment • humidifiers and dehumidifiers • air cleaners and filters.

Oil Burners fourth edition
Edwin M. Field
5 1/2 x 8 1/4   Hardcover   360 pp.   170 illus.
ISBN: 0-672-23394-0 $15.95

Up-to-date sourcebook on the construction, installation, operation, testing, servicing, and repair of all types of oil burners, both industrial and domestic: • general electrical hookup and wiring diagrams of automatic control systems • ignition system • high-voltage transportation • operational sequence of limit controls, thermostats, and various relays • combustion chambers • drafts • chimneys • drive couplings • fans or blowers • burner nozzles • fuel pumps.

Refrigeration: Home and Commercial second edition
Edwin P. Anderson; revised by Rex Miller
5 1/2 x 8 1/4   Hardcover   768 pp.   285 illus.
ISBN: 0-672-23396-7 $17.95

Practical, comprehensive reference for technicians, plant engineers, and homeowners on the installation, operation, servicing, and repair of everything from single refrigeration units to commercial and industrial systems: • refrigerants • compressors • thermoelectric cooling • service equipment and tools • cabinet maintenance and repairs • compressor lubrication systems • brine systems • supermarket and grocery refrigeration • locker plants • fans and blowers • piping • heat leakage • refrigeration-load calculations.

Hydraulics for Off-the-Road Equipment second edition
Harry L. Stewart; revised by Tom Philbin
5 1/2 x 8 1/4   Hardcover   256 pp.   175 illus.
ISBN: 0-8161-1701-2 $13.95

Complete reference manual for those who own and operate heavy equipment and for engineers, designers, installation and maintenance technicians, and shop mechanics: • hydraulic pumps, accumulators, and motors • force components • hydraulic control components • filters and filtration, lines and fittings, and fluids • hydrostatic transmissions • maintenance • troubleshooting.

Pneumatics and Hydraulics fourth edition
Harry L. Stewart; revised by Tom Philbin
5 1/2 x 8 1/4   Hardcover   512 pp.   315 illus.
ISBN: 0-672-23412-2 $15.95

Practical guide to the principles and applications of fluid power for engineers, designers, process planners, tool men, shop foremen, and mechanics: • pressure, work and power • general features of machines • hydraulic and pneumatic symbols • pressure boosters • air compressors and accessories • hydraulic power devices • hydraulic fluids • piping • air filters, pressure regulators, and lubricators • flow and pressure controls • pneumatic motors and tools • rotary hydraulic motors and hydraulic transmissions • pneumatic circuits • hydraulic circuits • servo systems.

Pumps fourth edition
Harry L. Stewart; revised by Tom Philbin
5 1/2 x 8 1/4   Hardcover   508 pp.   360 illus.
ISBN: 0-672-23400-9 $15.95

Comprehensive guide for operators, engineers, maintenance workers, inspectors, superintendents, and mechanics on principles and day-to-day operations of pumps: • centrifugal, rotary, reciprocating, and special service pumps • hydraulic accumulators • power transmission • hydraulic power tools • hydraulic cylinders • control valves • hydraulic fluids • fluid lines and fittings.

Carpenters and Builders Library
fifth edition 4 vols
John E. Ball; revised by Tom Philbin
5 1/2 x 8 1/4   Hardcover   1,224 pp.   1,010 illus.
ISBN: 0-672-23369-x $39.95
Also available in a new boxed set at no extra cost:
ISBN: 0-02-506450-9 $39.95

These profusely illustrated volumes, available in a handsome boxed edition, have set the professional standard for carpenters, joiners, and woodworkers.
Volume 1, Tools, Steel Square, Joinery
5 1/2 x 8 1/4   Hardcover   384 pp.   345 illus.
ISBN: 0-672-23365-7 $10.95

• Woods • nails • screws • bolts • the workbench • tools • using the steel square • joints and joinery • cabinetmaking joints • wood patternmaking • and kitchen cabinet construction.
Volume 2, Builders Math, Plans, Specifications
5 1/2 x 8 1/4   Hardcover   304 pp.   205 illus.
ISBN: 0-672-23366-5 $10.95

• Surveying • strength of timbers • practical drawing • architectural drawing • barn construction • small house construction • and home workshop layout.
Volume 3, Layouts, Foundations, Framing
5 1/2 x 8 1/4   Hardcover   272 pp.   215 illus.
ISBN: 0-672-23367-3 $10.95

• Foundations • concrete forms • concrete block construction • framing, girders and sills • skylights • porches and patios • chimneys, fireplaces, and stoves • insulation • solar energy and paneling.
Volume 4, Millwork, Power Tools, Painting
5 1/2 x 8 1/4   Hardcover   344 pp.   245 illus.
ISBN: 0-672-23368-1 $10.95

• Roofing, miter work • doors • windows, sheathing and siding • stairs • flooring • table saws, band saws, and jigsaws • wood lathes • sanders and combination tools • portable power tools • painting.

Complete Building Construction
second edition
John Phelps; revised by Tom Philbin
5 1/2 x 8 1/4   Hardcover   744 pp.   645 illus.
ISBN: 0-672-23377-0 $19.95

Comprehensive guide to constructing a frame or brick building from the footings to the ridge: • laying out building and excavation lines • making concrete forms and pouring fittings and foundation • making concrete slabs, walks, and driveways • laying concrete block, brick, and tile • building chimneys and fireplaces • framing, siding, and roofing • insulating • finishing the inside • building stairs • installing windows • hanging doors.

Complete Roofing Handbook

James E. Brumbaugh
5¹⁄₂ x 8¹⁄₄   Hardcover   536 pp.   510 illus.
ISBN: 0-02-517850-4 $29.95

Authoritative text and highly detailed drawings and photographs,on all aspects of roofing: • types of roofs • roofing and reroofing • roof and attic insulation and ventilation • skylights and roof openings • dormer construction • roof flashing details • shingles • roll roofing • built-up roofing • roofing with wood shingles and shakes • slate and tile roofing • installing gutters and downspouts • listings of professional and trade associations and roofing manufacturers.

Complete Siding Handbook

James E. Brumbaugh
5¹⁄₂ x 8¹⁄₄   Hardcover   512 pp.   450 illus.
ISBN: 0-02-517880-6 $23.95

Companion to *Complete Roofing Handbook*, with step-by-step instructions and drawings on every aspect of siding: • sidewalls and siding • wall preparation • wood board siding • plywood panel and lap siding • hardboard panel and lap siding • wood shingle and shake siding • aluminum and steel siding • vinyl siding • exterior paints and stains • refinishing of siding, gutter and downspout systems • listings of professional and trade associations and siding manufacturers.

Masons and Builders Library

second edition   2 vols
Louis M. Dezettel; revised by Tom Philbin
5¹⁄₂ x 8¹⁄₄   Hardcover   688 pp.   500 illus.
ISBN: 0-672-23401-7 $23.95

Two-volume set on practical instruction in all aspects of materials and methods of bricklaying and masonry: • brick • mortar • tools • bonding • corners, openings, and arches • chimneys and fireplaces • structural clay tile and glass block • brick walks, floors, and terraces • repair and maintenance • plasterboard and plaster • stone and rock masonry • reading blueprints.

Volume 1, Concrete, Block, Tile, Terrazzo
5¹⁄₂ x 8¹⁄₄   Hardcover   304 pp.   190 illus.
ISBN: 0-672-23402-5 $13.95

Volume 2, Bricklaying, Plastering, Rock Masonry, Clay Tile
5¹⁄₂ x 8¹⁄₄   Hardcover   384 pp.   310 illus.
ISBN: 0-672-23403-3 $12.95

Woodworking

Woodworking and Cabinetmaking

F. Richard Boller
5¹⁄₂ x 8¹⁄₄   Hardcover   360 pp.   455 illus.
ISBN: 0-02-512800-0 $16.95

Compact one-volume guide to the essentials of all aspects of woodworking: • properties of softwoods, hardwoods, plywood, and composition wood • design, function, appearance, and structure • project planning • hand tools • machines • portable electric tools • construction • the home workshop • and the projects themselves – stereo cabinet, speaker cabinets, bookcase, desk, platform bed, kitchen cabinets, bathroom vanity.

Wood Furniture: Finishing, Refinishing, Repairing second edition

James E. Brumbaugh
5¹⁄₂ x 8¹⁄₄   Hardcover   352 pp.   185 illus.
ISBN: 0-672-23409-2 $12.95

Complete, fully illustrated guide to repairing furniture and to finishing and refinishing wood surfaces for professional woodworkers and do-it-yourselfers: • tools and supplies • types of wood • veneering • inlaying • repairing, restoring, and stripping • wood preparation • staining • shellac, varnish, lacquer, paint and enamel, and oil and wax finishes • antiquing • gilding and bronzing • decorating furniture.

Maintenance and Repair

Building Maintenance second edition

Jules Oravetz
5¹⁄₂ x 8¹⁄₄   Hardcover   384 pp.   210 illus.
ISBN: 0-672-23278-2 $9.95

Complete information on professional maintenance procedures used in office, educational, and commercial buildings: • painting and decorating • plumbing and pipe fitting

• concrete and masonry • carpentry • roofing • glazing and caulking • sheet metal • electricity • air conditioning and refrigeration • insect and rodent control • heating • maintenance management • custodial practices.

Gardening, Landscaping and Grounds Maintenance

third edition
Jules Oravetz
5¹⁄₂ x 8¹⁄₄   Hardcover   424 pp.   340 illus.
ISBN: 0-672-23417-3 $15.95

Practical information for those who maintain lawns, gardens, and industrial, municipal, and estate grounds: • flowers, vegetables, berries, and house plants • greenhouses • lawns • hedges and vines • flowering shrubs and trees • shade, fruit and nut trees • evergreens • bird sanctuaries • fences • insect and rodent control • weed and brush control • roads, walks, and pavements • drainage • maintenance equipment • golf course planning and maintenance.

Home Maintenance and Repair: Walls, Ceilings and Floors

Gary D. Branson
8¹⁄₂ x 11 Paperback   80 pp.   80 illus.
ISBN: 0-672-23281-2 $6.95

Do-it-yourselfer's step-by-step guide to interior remodeling with professional results: • general maintenance • wallboard installation and repair • wallboard taping • plaster repair • texture paints • wallpaper techniques • paneling • sound control • ceiling tile • bath tile • energy conservation.

Painting and Decorating

Rex Miller and Glenn E. Baker
5¹⁄₂ x 8¹⁄₄   Hardcover   464 pp.   325 illus.
ISBN: 0-672-23405-x $18.95

Practical guide for painters, decorators, and homeowners to the most up-to-date materials and techniques: • job planning • tools and equipment needed • finishing materials • surface preparation • applying paint and stains · decorating with coverings • repairs and maintenance • color and decorating principles.

Tree Care second edition
John M. Haller
8½ x 11 Paperback 224 pp. 305 illus.
ISBN: 0-02-062870-6 $9.95

New edition of a standard in the field, for growers, nursery owners, foresters, landscapers, and homeowners: • planting • pruning • fertilizing • bracing and cabling • wound repair • grafting • spraying • disease and insect management • coping with environmental damage • removal • structure and physiology • recreational use.

Upholstering
updated
James E. Brumbaugh
5½ x 8¼ Hardcover 400 pp. 380 illus.
ISBN: 0-672-23372-x $12.95

Essentials of upholstering for professional, apprentice, and hobbyist: • furniture styles • tools and equipment • stripping • frame construction and repairs • finishing and refinishing wood surfaces • webbing • springs • burlap, stuffing, and muslin • pattern layout • cushions • foam padding • covers • channels and tufts • padded seats and slip seats • fabrics • plastics • furniture care.

Diesel Engine Manual fourth edition
Perry O. Black; revised by
William E. Scahill
5½ x 8¼ Hardcover 512 pp. 255 illus.
ISBN: 0-672-23371-1 $15.95

Detailed guide for mechanics, students, and others to all aspects of typical two- and four-cycle engines: • operating principles • fuel oil • diesel injection pumps • basic Mercedes diesels • diesel engine cylinders • lubrication • cooling systems • horsepower • engine-room procedures • diesel engine installation • automotive diesel engine • marine diesel engine • diesel electrical power plant • diesel engine service.

Gas Engine Manual third edition
Edwin P. Anderson; revised by
Charles G. Facklam
5½ x 8¼ Hardcover 424 pp. 225 illus.
ISBN: 0-8161-1707-1 $12.95

Indispensable sourcebook for those who operate, maintain, and repair gas engines of all types and sizes: • fundamentals and classifications of engines · engine parts • pistons • crankshafts • valves • lubrication, cooling, fuel, ignition, emission

control and electrical systems • engine tune-up • servicing of pistons and piston rings, cylinder blocks, connecting rods and crankshafts, valves and valve gears, carburetors, and electrical systems.

Small Gasoline Engines
Rex Miller and Mark Richard Miller
5½ x 8¼ Hardcover 640 pp. 525 illus.
ISBN: 0-672-23414-9 $16.95

Practical information for those who repair, maintain, and overhaul two- and four-cycle engines – with emphasis on one-cylinder motors – including lawn mowers, edgers, grass sweepers, snowblowers, emergency electrical generators, outboard motors, and other equipment up to ten horsepower: • carburetors, emission controls, and ignition systems • starting systems • hand tools • safety • power generation • engine operations • lubrication systems • power drivers • preventive maintenance • step-by-step overhauling procedures • troubleshooting • testing and inspection • cylinder block servicing.

Truck Guide Library 3 vols
James E. Brumbaugh
5½ x 8¼ Hardcover 2,144 pp. 1,715 illus.
ISBN: 0-672-23392-4 $45.95

Three-volume comprehensive and profusely illustrated reference on truck operation and maintenance.

Volume 1, Engines
5½ x 8¼ Hardcover 416 pp. 290 illus.
ISBN: 0-672-23356-8 $16.95

• Basic components · engine operating principles • troubleshooting • cylinder blocks • connecting rods, pistons, and rings • crankshafts, main bearings, and flywheels • camshafts and valve trains • engine valves.

Volume 2, Engine Auxiliary Systems
5½ x 8¼ Hardcover 704 pp. 520 illus.
ISBN: 0-672-23357-6 $16.95

• Battery and electrical systems • spark plugs • ignition systems, charging and starting systems • lubricating, cooling, and fuel systems • carburetors and governors • diesel systems • exhaust and emission-control systems.

Volume 3, Transmissions, Steering, and Brakes
5½ x 8¼ Hardcover 1,024 pp. 905 illus.
ISBN: 0-672-23406-0 $16.95

• Clutches • manual, auxiliary, and automatic transmissions • frame and suspension systems • differentials and axles, manual and power steering • front-end alignment • hydraulic, power, and air brakes • wheels and tires • trailers.

Answers on Blueprint Reading
fourth edition
Roland E. Palmquist; revised by
Thomas J. Morrisey
5½ x 8¼ Hardcover 320 pp. 275 illus.
ISBN: 0-8161-1704-7 $12.95

Complete question-and-answer instruction manual on blueprints of machines and tools, electrical systems, and architecture: • drafting scale • drafting instruments • conventional lines and representations • pictorial drawings • geometry of drafting • orthographic and working drawings • surfaces • detail drawing • sketching • map and topographical drawings • graphic symbols • architectural drawings • electrical blueprints • computer-aided design and drafting. Also included is an appendix of measurements • metric conversions • screw threads and tap drill sizes • number and letter sizes of drills with decimal equivalents • double depth of threads • tapers and angles.

Complete Course in Stained Glass
Pepe Mendez
8½ x 11 Paperback 80 pp. 50 illus.
ISBN: 0-672-23287-1 $8.95

Guide to the tools, materials, and techniques of the art of stained glass, with ten fully illustrated lessons: • how to cut glass • cartoon and pattern drawing • assembling and cementing • making lamps using various techniques • electrical components for completing lamps • sources of materials • glossary of terminology and techniques of stained glasswork.

Macmillan Practical Arts Library
Books for and by the Craftsman

World Woods in Color

W.A. Lincoln
7 × 10   Hardcover   300 pages
300 photos
ISBN: 0-02-572350-2 $39.95

Large full-color photographs show the natural grain and features of nearly 300 woods: • commercial and botanical names • physical characteristics, mechanical properties, seasoning, working properties, durability, and uses • the height, diameter, bark, and places of distribution of each tree • indexing of botanical, trade, commercial, local, and family names • a full bibliography of publications on timber study and identification.

The Woodturner's Art: Fundamentals and Projects

Ron Roszkiewicz
8 × 10   Hardcover   256 pages   300 illus.
ISBN: 0-02-605250-4 $24.95

A master woodturner shows how to design and create increasingly difficult projects step-by-step in this book suitable for the beginner and the more advanced student: • spindle and faceplate turning • tools • techniques • classic turnings from various historical periods • more than 30 types of projects including boxes, furniture, vases, and candlesticks • making duplicates • projects using combinations of techniques and more than one kind of wood. Author has also written *The Woodturner's Companion*.

The Woodworker's Bible

Alf Martensson
8 × 10   Paperback   288 pages   900 illus.
ISBN: 0-02-011940-2 $12.95

For the craftsperson familiar with basic carpentry skills, a guide to creating professional-quality furniture, cabinetry, and objects d'art in the home workshop: • techniques and expert advice on fine craftsmanship whether tooled by hand or machine • joint-making • assembling to ensure fit • finishes. Author, who lives in London and runs a workshop called Woodstock, has also written. *The Book of Furnituremaking.*

Cabinetmaking and Millwork

John L. Feirer
7 1/8 × 9 1/2   Hardcover   992 pages
2,350 illus. (32 pp. in color)
ISBN: 0-02-537350-1 $47.50

The classic on cabinetmaking that covers in detail all of the materials, tools, machines, and processes used in building cabinets and interiors, the production of furniture, and other work of the finish carpenter and millwright: • fixed installations such as paneling, built-ins, and cabinets • movable wood products such as furniture and fixtures • which woods to use, and why and how to use them in the interiors of homes and commercial buildings • metrics and plastics in furniture construction.

Cabinetmaking: The Professional Approach

Alan Peters
8 1/2 × 11   Hardcover   208 pages   175 illus.
(8 pp. color)
ISBN: 0-02-596200-0 $29.95

A unique guide to all aspects of professional furniture making, from an English master craftsman: • the Cotswold School and the birth of the furniture movement • setting up a professional shop • equipment • finance and business efficiency • furniture design • working to commission • batch production, training, and techniques • plans for nine projects.

Carpentry and Building Construction

John L. Feirer and Gilbert R. Hutchings
7 1/2 × 9 1/2   hardcover   1,120 pages
2,000 photos (8 pp. in color)
ISBN: 0-02-537360-9 $50.00

A classic by Feirer on each detail of modern construction: • the various machines, tools, and equipment from which the builder can choose • laying of a foundation • building frames for each part of a building • details of interior and exterior work • painting and finishing • reading plans • chimneys and fireplaces • ventilation • assembling prefabricated houses.